The
Sorcerers
and Their
Apprentices

The Sorcerers

and Their Apprentices

How the Digital Magicians of the
MIT Media Lab Are Creating the
Innovative Technologies That Will
Transform Our Lives

Frank Moss

Former Director of the MIT Media Lab

CROWN
BUSINESS
NEW YORK

CROWN BUSINESS is a trademark, and CROWN and the Rising Sun colophon are
registered trademarks of Random House, Inc.

Library of Congress Cataloging-in-Publication Data

Moss, Frank, 1949–
The sorcerers and their apprentices : how the digital magicians of the MIT Media Lab
are creating the innovative technologies that will transform our lives / Frank Moss.
p.   cm.
Summary: "From the former director of the famed MIT Media Laboratory comes an
exhilarating behind-the-scenes exploration of the research center where our nation's foremost
scientists are creating the innovative new technologies that will transform our future"—
Provided by publisher.
Includes index.
1. Massachusetts Institute of Technology. Media Laboratory. 2. Digital communications—
Research—United States. 3. Scientists—Massachusetts—Cambridge—Intellectual
life—21st century. I. Title.

TK5103.7.M675 2011   607.2'7444—dc22

2010052742

ISBN 978-0-307-58910-1
eISBN 978-0-307-58912-5

Printed in the United States of America

Book design by Maria Elias
Jacket photograph by Sam Ogden

2   4   6   8   10   9   7   5   3   1

First Edition

To Professor Seymour Papert,

whose passion for children and learning appended "heart" to
"mind and hand" in the MIT motto . . .

and who changed my entire perspective on the future of people
and technology during a fifteen-minute chat on Cape Cod

Contents

Preface

"We must humanize technology before it dehumanizes us."
—Oliver Sacks, neurologist and author[*]

In December of 2005 I was struggling with a decision. Should I accept an offer to become the MIT Media Lab's next director? It looked to be a dream job for a career technologist and entrepreneur such as myself, and it fit well with my desire to contribute more directly to the challenges facing society. However, I was wondering if this move made any sense at all for me, given the well-known cultural differences between business and academia that I had heard so much about.

To get an insider's view, I called Professor Rodney Brooks, then director of MIT's Computer Science and Artificial Intelligence Laboratory (CSAIL). Brooks is a distinguished academic in the field of robotics and also a successful entrepreneur, so I guessed he might relate to my dilemma. When I asked why he had made the decision to become

---

[*] Keynote Speech, MIT Media Lab symposium entitled "h2.0 New Minds, New Bodies, New Identities," May 9, 2007, http://h20.media.mit.edu/.

director of CSAIL, he pondered the question for a moment and then declared, "Besides the opportunity to work with the smartest people anywhere, the biggest advantage of being director of a major MIT lab is that if there is a story you want to tell, it's a fantastic stage from which to tell it."

After I took the job, I thought about this often, and before long the story I wanted to tell became clear. Actually it was many stories. Perhaps it's not so surprising that as I traveled around talking about the Media Lab, audiences were blown away by the latest inventions that I showcased. But what struck me was that they were equally impressed by the stories I told about how these inventions actually came into being—how they grew out of an environment that totally defies the conventional wisdom of what a research lab "should be." At the Media Lab, people from an astonishing variety of disciplines (architects, computer scientists, electrical engineers, musicians, neuroscientists, physicists, visual artists, just to name a few) enjoy the unrestricted freedom to create and invent as their passions dictate. They don't sit around and wait for inspiration to hit. Rather than thinking about what to build, they dive in and build what they are thinking about. This unfolds totally out in the open, in cluttered workshops where students rub shoulders regularly with liaisons from the world's largest corporations. The only guidelines for the researchers are that (1) their inventions have the potential to significantly improve people's lives in the future and (2) their work is radically different from what anyone else is doing.

In this open and anything-goes environment, ideas germinate, cross-pollinate, and mutate in a fashion that is literally out of control. But that's the idea. Out this creative chaos emerge literally hundreds of inventions a year, from the practical to the lunatic. It is not uncommon for one or more of them to survive and somehow grow into an innovation that disrupts industries, spawns entirely new ones, or even transforms society.

Inspired by the enthusiastic reception of my audiences, I have designed this book as a collection of stories about the faculty and student inventors at the Media Lab (the "sorcerers and their apprentices"), providing a behind-the-scenes look into the fascinating process (the "magic") by which they create and invent. The stories that I have elected to tell, roughly two dozen in all, cover only a fraction of the inventions that are generated at the Media Lab every year. However, these are ones that I find to be particularly representative of both the Media Lab's highly unorthodox approach to invention and innovation as well as the exciting new directions that its research is taking today.

I'm not the first to write a book about the inventors and their inventions at the MIT Media Lab. Shortly after it officially opened its doors in 1985, the futurist Stewart Brand introduced it to the world in his 1987 volume *The Media Lab: Inventing the Future at MIT.* In it, Brand, who had spent some time in residence at the fledgling Media Lab, described the efforts of the early team, most of who didn't fit into the traditional academic silos of MIT but thrived in the eclectic, highly unorthodox environment of the Media Lab. This included cofounder and first director Nicholas Negroponte, who famously predicted the "digital convergence" of three industries—print/publishing, broadcast/entertainment, and computers—as well as the seismic shifts that this convergence would have on people, industries, and society. Negroponte later elaborated on these in his own 1995 book *Being Digital.*

While I was in the process of completing this book, the Media Lab celebrated its twenty-fifth anniversary. By this time, in fall of 2010, the digital convergence that Negroponte predicted had indeed occurred. Today's social, mobile, aware, real-time, and hyperconnected world of people and information, which the Media Lab helped to catalyze, has changed how we live, work, and play to a degree perhaps even greater than Negroponte or his cofounders could have possibly imagined.

But as amazing as these technologies seem to us, they are mere "digital affordances" compared to what is being imagined and invented at the Media Lab today. In addition to tapping into the full power of information and communication technologies, researchers at the Lab are also leveraging equally dramatic advances in the biological, physical, and social sciences, enabling them to create a new generation of inventions that will have a much deeper impact on people's lives. Heeding Oliver Sacks's urgent entreaty to "humanize technology," the Media Lab's mission for the next quarter century is to *empower ordinary people to do truly extraordinary things and, in the process, take control over the most important aspects of their lives—their health, their wealth, and their happiness.*

In the pages ahead, you will read about revolutionary technologies being developed in the workshops at the Media Lab today that:

- Are much simpler and much less intrusive in people's everyday lives
- Augment human mental and physical abilities, beginning with the "disabled"
- Learn from people, understand them, and are highly responsive to their wants and needs
- Help people reflect and act on their "life data" to make truly rational decisions

- Make homes, workplaces, and cities adaptable to their human inhabitants
- Unleash the full creative powers within every human being

I would like to offer the reader just a brief explanation of how I organized the stories in this book and explain some key messages that I would like to emphasize. Like the Media Lab itself, this book doesn't have a very formal and rigid structure. But roughly, each chapter contains about three stories about the inventors and their inventions, which taken together help illustrate a particular theme.

In chapters one through four, the theme of each is a fundamental principle that underlies the Media Lab's unique approach to innovation, as follows:

- Chapter 1, "The Power of Passion," introduces the unprecedented *creative freedom* that its researchers enjoy to invent according to their passions and curiosities, in an environment where the only real rule is that there are no rules, and where there is no such thing as a failure.
- Chapter 2, "Disappearing Disciplines," describes the Lab's *anti-disciplinary* ethos, where people from widely different backgrounds think about problems in wildly different ways from the past, unencumbered by preconceived notions of what is possible or what the solutions "should" look like.
- Chapter 3, "Hard Fun," is about the Media Lab's distinctive approach to *playful invention*, which begins by teaching students how to build almost anything and then encourages them to express their most fanciful ideas by building them and then seeing what happens when people use them.

- Chapter 4, "Serendipity by Design," explains how the Media Lab deliberately fosters an environment in which the kinds of unlikely and seemingly random connections that spark truly big ideas not only happen but *can't help but happen.*

While I must admit that the MIT Media Lab is a very special place, I believe that these four principles, either in whole or in part, can be adopted by any individual, business, or institution—from start-ups to multinational corporations, from toy companies to financial services firms, from schools to government agencies—to improve their own process of innovation.

Moreover, it is my hope that these four principles can inform a desperately needed rethinking of our innovation ecosystem in the United States. There is near universal agreement that innovation is the key to confronting the urgent challenges that face humanity in the twenty-first century. But tragically, the fifty-year-old "deal" between government, industry, and academia that spawned the wave of innovation in the United States in the late twentieth century is broken. Government agencies, whose visionary leaders and money were the forces behind transformational advances such as the Internet and the mapping of the human genome, today support baby steps rather than bold leaps. University researchers play it safe in their grant proposals, aware that it is unlikely that peer reviewers will approve radical or controversial ideas. Large companies (with a few exceptions, such as Google) have dramatically cut their investments in curiosity-driven research, and most venture capitalists, whose long-shot bets fueled the high-tech and biotech revolutions, have lost their taste for early-stage investments. The result is that radical new ideas and inventions, which are the seeds of innovation, are no longer being created at nearly the pace they were before. I believe that the way they are generated in abundance at the Media Lab, as described in this book, can serve as a guide to help us reverse that disastrous path.

The stories continue in chapters five through eight, but the themes shift to focus on the fundamental ways in which the technologies under development at the Media Lab today will transform our lives, society, and business in the future.

- Chapter 5, "The New Normal," presents *human-augmentation* technologies that will forever alter our most basic concepts of human abilities, first addressing the challenges of people normally considered to be disabled, such as amputees and people with autism, but ultimately improving the lives of everyone.
- Chapter 6, "Living and Learning Together," explores a new relationship between people and technology, in which technology learns from, understands, and helps people as true partners, from smart phones that serve as savvy personal assistants to sociable robots that provide aid and companionship to the elderly.
- Chapter 7, "The Age of Agency," looks at how technology will eliminate the age-old asymmetry between ordinary people and the "high priests" of society, such as doctors and bankers, empowering individuals with unprecedented control over their health and finances.
- Chapter 8, "I Am a Creator," is about technologies that unleash the full powers of expression and creativity that exist within each and

every human being, just waiting to be released, and how these will transform the very identity of individuals and society as a whole in the future.

I would like to offer one final thought. After nearly thirty years as a technologist in the business of innovation, the five years that I have spent at the Media Lab have completely changed my view of the possibilities that technologic innovation holds for confronting the challenges we face as people and society. I am now convinced that *individuals,* empowered with the type of radically new technologies that you will read about in the pages ahead, can succeed in transforming society from the bottom up where our *institutions* have dismally failed. This has greatly increased my sense of optimism for the future, and I hope it will do the same for you.

The
Sorcerers
and Their
Apprentices

CHAPTER 1

The Power of Passion

"Andy, are we ready?"

Grant Elliott, a twenty-eight-year-old PhD candidate in electrical engineering, is hovering over Andrew Marecki, a twenty-one-year-old MIT senior mechanical engineering major, who is working with him for the summer. Marecki is sitting cross-legged on the floor of the crowded *Biomechatronics* workshop. Scattered before them on the ground are two long, thin fiberglass rods called *struts,* assorted pieces of hardware, a tangle of black Velcro straps, and a pair of black knit wetsuit shorts. To the casual observer, it looks like nothing but a pile of junk, waiting to be hauled away with the morning's trash. But to these young inventors, these are the ingredients of a new—and ambitious—challenge.

Every surface of the workshop is cluttered with computers, scraps of paper, assorted tools, and miscellaneous parts. Unidentifiable objects dangle from hooks and straps suspended from the ceiling. The Biomechatronics workshop and the several dozen other research groups that occupy the Wiesner Building in the summer of 2009 are bursting at their seams. In a few months, this structure, which has been home to the Media Lab for the past twenty-five years, will be united with the long-awaited Media Lab extension: a six-story, shimmering aluminum and glass digital wonderland designed by the great Japanese architect Fumihiko Maki.

At arm's length from where Elliott and Marecki are hard at work is a red table strewn with disembodied mechanical limbs: a piece of a prosthetic knee and the vestiges of a robotic foot that had been taken apart, its pieces stripped and cannibalized for other projects. On the opposite counter sits a turquoise sewing machine that looks oddly out of place—more at home, perhaps, on the set of a fashion show than on the counter in a workshop at one of the most prestigious research labs in the world. It was in fact used the night before to repair the wetsuit shorts that are part of the mess that Marecki is trying to untangle. The shorts had gotten torn during an experiment on the previous day.

Biomechatronics is an exciting new field that lies at the intersection of biology and engineering. It involves measuring exactly how people walk, climb, and run and then transforming this information into sophisticated computer models of human locomotion. It then uses those models to create *biologically inspired prostheses* for amputees—and more recently, *augmentative exoskeletons* for the able bodied—that meld seamlessly with the body. The ultimate goal of biomechatronics is to create smart electromechanical replacements that restore physiological function in humans: devices that interface so naturally with the human nervous system that they feel and behave like the real thing.

We have yet to build a prosthetic foot that allows the wearer to feel the blades of grass beneath his or her feet or the grains of sand on a beach. However, the Biomechatronics group—which, in addition to

Marecki and Elliott, consists of two mechanical engineers, two physicists, two machine learning experts, a material science expert, and an electrical engineer—is inching closer toward that goal every day. The team works at a dizzying pace, and although each member is assigned a different task, the reality is that everyone ends up doing everything. As Elliott, who has both a bachelor's degree in physics and a master's degree in electrical engineering, notes, "To work in this group, you have to be a bit of a Renaissance person."

This is typical of the highly interdisciplinary nature of the Media Lab, a cauldron of creativity and invention where there are no boundaries between fields or disciplines and no one is confined to his or her specialty. This is a place where computer scientists study design and early childhood education, where musicians conduct research in neuroscience, where artists become proficient in electrical engineering and building robots, and where dreamers and thinkers become doers and inventors. The Media Lab's interdisciplinary approach is also reflected in the scope and breadth of our twenty-five different research groups, whose names reflect their eclectic and completely original work. Among them are *Personal Robots, Opera of the Future, Lifelong Kindergarten, New Media Medicine, Affective Computing, Viral Communications, Cognitive Machines, Speech + Mobility, Information Ecology* and *Tangible Media.*

The faculty and students—the sorcerers and their apprentices—who make up these groups are truly exceptional individuals who would accomplish great things anywhere. But when you put them in the environment of the Media Lab, their passions are unleashed to the fullest. Here they are able to pursue their visions with unmatched creative freedom, taking bold risks that would be unthinkable elsewhere. Here they can fearlessly cross boundaries between traditional disciplines, reframing old problems in surprising new ways. Here they can build just about anything they can imagine, and they can observe what happens when people use their inventions in the real world. Here they can turn setbacks into learning experiences and stepping stones to valuable new insights and ideas. And they can do all this thanks to the unique and highly unorthodox approach to invention and innovation that has evolved at the MIT Media Lab over a quarter of a century.

THE ORIGINAL PROTOTYPE

The Media Lab, known for its focus on building and prototyping, actually began its own life as a prototype of sorts. Cofounded in 1985 by two visionary technologists and humanists—Jerome Wiesner, then recently retired as MIT president, and Nicholas Negroponte, then a young faculty member in the MIT Department of Architecture—the Lab was conceived as a place to study and prepare for the coming digital revolution by breaking down the barriers between the isolated "silos" of disciplines that were a fixture of twentieth-century academia. Decades before smart phones and e-books became staples of our everyday lives, Wiesner and Negroponte predicted that rapid advances in information and communication technologies would soon spark a three-way *media convergence* of computers, newspapers, and television and that this would have a highly disruptive impact on both society and business. So with Negroponte as the Lab's first director, together they recruited a number of like-minded MIT colleagues from a wide variety of backgrounds, most of whom didn't fit well with the traditional academic structure of the institute at that time. But the question still remained: How would they fund all the ambitious research they had in mind? Soon, they had their answer. In a highly unorthodox move they convinced a number of large companies—mainly in the computer, telecommunications, media, and financial services industries—to enter into an unprecedented financial and intellectual collaboration. The companies would fund the operations of the Media Lab in return for an equal but nonexclusive share in the resulting intellectual property the researchers generated, with very few strings attached. The result was that the researchers enjoyed a degree of creative freedom that was unthinkable elsewhere.

Next, the pair also quickly raised the funds for a new facility that, like its researchers, was distinctly different from anything on the MIT campus. The new building, named in Wiesner's honor, was designed by I. M. Pei, and it included a two-story performance space, "computer gardens," open areas, and an informal atmosphere where research was ongoing twenty-four hours a day. The machinery to make the place run—including the latest experimental communications networks—was hidden in floors, closets, and behind doors. The excitement and energy in this new and very different place was palpable to anyone who wandered through its spaces. The Lab grew at a dizzying pace not just in its numbers of faculty and students but also in the variety of disciplines they represented: visual arts and design, the computer sciences, the physical sciences, mechanical and electrical engineering, electronic music, epistemology and learning, and more. It was an eclectic crew, uniquely poised to be on the cutting edge of the rapidly evolving relationship between people and technology.

From the very outset, the Media Lab enjoyed a degree of independence that is extremely rare in the academic world. While all other labs at MIT drew their faculty from a variety of academic departments, thanks to an unprecedented agreement made by its founders with the MIT administration, the Media Lab contained its own academic program, called "Media Arts and Sciences" (MAS). For historical reasons, MAS technically fell within the MIT School of Architecture and Planning, but the Lab was less like a formal department and more like an interdisciplinary school within a school. MAS could appoint its own faculty and select its own students, who were working toward master of science and PhD degrees. In time the Media Lab grew to house about 150 graduate students divided up among more than two dozen groups, each headed by a member of the Media Arts and Sciences faculty or Media Lab research staff. This close coupling between the academic program and research was a tremendous advantage. Since many of the required courses in MAS were actually hands-on opportunities to invent and create, this arrangement resulted in an unprecedented production of new ideas and inventions.

But equally important to the Media Lab's unique model for innovation was the unique partnership that developed between the researchers and corporate sponsors. In another example of the Media Lab's aversion toward boundaries of any kind, employees from sponsor companies were encouraged to collaborate closely with the faculty and students. Some employees even took sabbaticals from their own jobs to become full members of a Lab research team for a year or longer. At the bare minimum, researchers and sponsor representatives would come together twice a year at weeklong sponsor meetings, during which students, in a ritual that soon came to be known as "demo or die," rolled out and demonstrated all of their latest creations. The Media Lab became famous for these extravaganzas—known as "hell week" because highly caffeinated students labored furiously through the days and nights preceding the event in order to get their projects ready to show to sponsors—and they are still a huge part of Media Lab culture today.

Before long the Media Lab became the destination for anyone from anywhere in the world who wished to experience its uniquely creative environment and who wanted a glimpse into the future impact of technology on society, business, and everyday life. In fact, over the years the Media Lab has contributed in many ways to the digital lifestyle that we enjoy today. If you've ever read a book on an Amazon Kindle, let Guitar Hero unleash your inner rock star, built a LEGO Mindstorms robot, downloaded a movie or music file, or driven in a vehicle with child-safe airbags, you have enjoyed the benefits of technology dreamed up at the Media Lab.

In the same way that the myriad technologies developed at the Lab evolve through a series of iterations and progressions, so has the Lab itself evolved over these past two and a half decades. The "media convergence" that Negroponte and Wiesner predicted a quarter century ago has indeed taken place, perhaps to an extent greater than even they could have imagined. But the Lab's mission is far from complete. What most people don't realize is that the Media Lab has become about so much more than digital gizmos and gadgets and occasional lunatic creations. Today it's about harnessing the full power of computers, networks, and other emerging technologies to improve the quality of our lives at a much deeper level, to address the many pressing problems confronting society and business in the twenty-first century, and to create a better future for everyone.

As you'll read about in the chapters ahead, projects currently under way at the Lab include CityCar, a networked, digitally controlled, stackable, foldable electric vehicle that will make our urban areas much more livable, safe, and sustainable; CollaboRhythm, a software platform that uses virtual personal health care assistants to enable patients take control of their own health at home or anywhere; FaceSense, a small wearable device that tracks and translates facial expressions into emotions in real time, helping people with autism to function better in school and society; Nexi, one of the world's first mobile, dexterous, and sociable robots, a robot so sophisticated it can learn from and live together with elderly people, providing much-needed companionship and assistance; Merry Miser, a personalized smart phone app that empowers people to take control of their own finances by helping them make more informed spending decisions while they shop; and Mobile Musical Diagnostics, which let individuals use their playlists to detect memory disorders in the earliest stage.

MY JOURNEY TO THE CENTER OF INNOVATION

I became the third director of the MIT Media Lab in early 2006, taking over from then acting director Walter Bender, who stepped in when Nicholas Negroponte left to pursue his One Laptop Per Child (OLPC) project aimed at providing low-cost laptops for children in the developing world. My journey to the Media Lab involved a great deal of serendipity, much as many of the Lab's inventions. It actually began at MIT in the early 1970s, nearly a decade before the Lab opened its doors. Intent on following my childhood dream of working in the space program, I enrolled as a graduate student in MIT's Department of Aeronautics and Astronautics. But by the time I earned my doctorate in 1976, the moon race was over. The public had lost its fascination with space exploration, and as a result government funding for it had been slashed—even the scientists who had designed *Apollo* rockets were out of jobs. Clearly I needed to take my career in a different direction. Fortunately, I had seen the writing on the wall and had already begun to move my doctoral research away from aeronautics and toward a new field of study that was just then starting to emerge: networks of computers. Like many others at MIT at the time, I was intrigued by the vast potential that emerging computer and communications technologies held to transform society, and I wanted to be a part of it. The computer industry seemed like a natural "backup plan."

So I took a job as a network researcher in the computer sciences department of IBM's T.J. Watson Research Lab in Yorktown Heights, New York. But after six years I became impatient. I wanted to see my ideas translated into practical use much faster than was possible at IBM at that time. I left IBM in 1984 and moved to the Boston area, which was then in the thick of the "Route 128" high-tech boom, to try my hand in the entrepreneurial world. After stints as a software vice president at Apollo Computer, Stellar Computer, and Lotus Development, in 1991 I moved with my wife and youngest daughter to Austin, Texas, where I became the CEO of Tivoli Systems, an "upstart" startup that was determined to challenge the reigning technology giants in the exploding market for software that managed and controlled large computer networks. In 1995 I took Tivoli public. It then merged with IBM a little more than a year later (in what was called a *reverse merger*) and I stayed on for several years as chairman of the Tivoli software business unit. When I left IBM in 1998, now for the second time in my career, I felt that entrepreneurial itch once again. It was the height of the dot-com boom, and I couldn't resist jumping back into the startup game. So I moved with my wife and daughter back to Boston—we wanted to be closer to our families again—and I cofounded a couple of Internet companies.

But after nearly twenty-five years in the computer game, I wasn't enjoying it nearly as much as I had before. I had become disenchanted by what I perceived as the limited impact that computers and networks had actually had on the quality of our lives—that is, limited relative to the high hopes I'd had when I first got into the industry a quarter century before. For all the hoopla of the information revolution, I felt that my industry wasn't producing the kinds of transformative innovations that I'd envisioned. In my view, we were creating mere *digital affordances*: technologies that changed the way people consume information, shop, and socialize but that didn't address those really big challenges like climate change, poverty, and chronic disease facing us as individuals and society. Just building another high-tech startup company only to sell it or take it public no longer excited me. I wanted to do what I'd always urged my children to do and, ironically, what they were now urging me to do: Find a way to devote my energies to something not only that I might really enjoy but something that also might help make the world a better place.

I had started to become intrigued by the growing role of information technology in biomedical research, so I reached out to some prominent players in the Boston scientific community to see if I could somehow get involved in this coming revolution in human health. I was surprised by their enthusiasm. The activity in this area was increasingly becoming data driven, and my background in the computer business was apparently an asset. The late Alex d'Arbeloff, an acquaintance from the high-tech world and then chairman of the MIT Corporation, introduced me to MIT professor of biology Eric Lander, who was then in the final phases of the ambitious effort to sequence the human genome. Lander invited me to join a small group advising him on creating a new biomedical research center in Cambridge devoted to leveraging the power of genomics to revolutionize medicine. This is where I met Steven Holtzman, a highly regarded veteran of the biotechnology industry, who at that moment in time was working on a new venture focused on finding novel approaches to cancer drug discovery. Holtzman and I became fast friends and over coffees brainstormed about how information technology might help accelerate the new company's search for new medicines. When Holtzman invited me to become a cofounder of Infinity Pharmaceuticals, I immediately accepted. Suddenly, I was no longer just in the computer business. I was in the business of using computers to facilitate the process of discovering important new medicines. I was also having more fun than I'd had in years.

Then, out of the blue, on a November afternoon in 2005, I received a call from an executive search firm in Boston. Was I interested, the headhunter asked, in being a candidate for a new job opening at MIT, as the director of the Media Lab? This was really odd, I remember thinking. Why would they want me? I wasn't an academic. I was a career entrepreneur. I asked how he had gotten my name, and he said that Steve Holtzman had referred me to him. I thought this was very nice of Steve, but it held little interest for me. Sure, I'd heard about the Media

Lab over the years, as it had become famous in high-tech circles, and I had even visited once in the late eighties. But I knew almost nothing about what they did there, and my impression was that by now it was mostly famous for being famous. Plus, why at this point in my career would I choose to enter academia? That wouldn't be for me, I thought.

I politely declined the invitation to visit, but the headhunter wouldn't take no for an answer. For some reason he felt strongly that this was a perfect fit. After a few more weeks of persistent e-mails and voice mails from him, I finally agreed to spend a day at the lab. But I was still very skeptical. I expected to find a bunch of computer whizzes tinkering over obscure and esoteric inventions—people stifled by that same narrow focus that had sent me running from the high-tech world five years earlier.

I couldn't have been more wrong. Instead, I discovered a building full of passionate, wildly energetic people who were beginning to move technology in a new, different, and decidedly better direction. These were people with an incredibly diverse set of interests who, instead of devoting their time and energies to mere digital affordances, were inventing things that would truly make a difference in people's lives. The things I saw on that freezing December afternoon didn't just alter my career trajectory. They reshaped the way I thought about creativity and innovation, forever.

Professor Tod Machover, who was the faculty member leading the search for the new director, began my tour in his *Opera of the Future* workshop with its bare-to-the-bone exposed ceiling and open ductwork. There he introduced me to one of his students, Adam Boulanger, who was busy tinkering with his latest invention: an infrared head tracker that enabled people with severe physical disabilities to compose music using a software tool the group had developed called Hyperscore. He told me that his goal was to help people unleash their creative and expressive powers in a way no one ever thought possible. As I was soon to learn, this was the objective of many other projects going on at the lab. I remember being struck by the ease with which Boulanger task switched between working away on his prototype and delivering an articulate and impassioned explanation of how it worked.

As the afternoon, and my tour, went on, more surprises began to emerge. Instead of finding people just staring at computer screens, I also found them huddled over worktables whittling, drilling, drawing, soldering, wiring, sewing, and sculpting. Machover explained to me that the Media Lab wasn't just a place where you dream up inventions. Here, you are expected to actually build, test, and demonstrate them. This approach was inspired, he added, by his beloved friend and colleague Professor Seymour Papert, who believed that creativity can best be unleashed by taking the hands-on approach to learning and building that comes naturally to curious children at play. As I learned that day, this idea—what Papert called the art of "Hard Fun"—is central to the Media Lab's unique approach to invention and innovation.

As we continued through the building, I took the opportunity to chat briefly with some of the students. I remember being particularly impressed by the incredible diversity of their backgrounds and interests. Instead of the homogenous band of engineers and computer scientists I had expected to meet, I encountered people who had studied and worked in fields ranging from the visual arts to psychology to poetry to music. No one was confined to his or her "specialty"; instead, computer scientists were exploring the psychology of early childhood learning, musicians were developing computer programming and were collaborating with neuroscientists, and artists were proficient in electrical engineering and built humanoid robots.

The picture became even clearer when Professor Pattie Maes, a native of Belgium and a leader in the field of human-computer interaction (HCI), told me how the style of research at the Media Lab—what she called the "atelier" method—was a modern adaptation of the centuries-old apprenticeship model. Creative geniuses like da Vinci, Michelangelo, and Copernicus and other masters of their craft would employ younger helpers, paying them in the form of housing, food, and formal instruction in the craft. While working side by side, sometimes for years on end, the masters would pass their skills and knowledge on to their protégés, seeding a new generation of inventors. At the Media Lab, Maes explained, faculty members are like those master craftsmen, and the students, who learn their "craft" through years of hands-on research, are like their apprentices. Next, Maes showed me around her Ambient Intelligence workshop, located in an open area of the lab she referred to as "the pond." The walls, tables, and floors were covered with dozens of inventions created by her group, some of which I recognized from reading about them in the press over the years. Several of her students were engrossed in their work, but when she asked one of them, a PhD student named David Merrill, to give me a quick demo of his project, he readily agreed. Merrill walked us over to a three-foot-wide mockup of a supermarket shelf stocked with cartons of butter, Egg Beaters, and cereal, and he happily slipped on a Bluetooth-enabled ring he had been tinkering with when we interrupted him. He pointed directly at a box of cereal, and a light on the shelf directly below it glowed red. This meant, he told us, that the food didn't fit the nutritional profile that he had programmed into the device. Perhaps it contained nuts or not enough fiber.

He told me that there were a lot of "really cool technologies" making this happen—an infrared transmitter/receiver mounted on the ring, a transponder on the shelf with which it communicated, and a Bluetooth connection to a smart phone that could access the wearer's profile in real time, to name a few. It was easy to see how this "augmented reality interface," as Merrill called it, could change the experience of in-store shopping in truly a profound way. But what really impressed me during this visit was the close working relationship he clearly enjoyed with Maes. He called her "Pattie," and my impression was that they engaged in give-and-take like true collaborators and colleagues.

The day had begun as a courtesy visit, but now my interest was piqued. I was particularly struck by the incredible passion I encountered among the professors and the students I'd met thus far and by how committed they were to finding radical new ways for technology to improve the quality of life for everyone, including the disabled and disadvantaged of society. What's more, I was amazed by the remarkable creative freedom they enjoyed to break down traditional barriers and pursue these passions together. This wasn't at all what I had expected. However, I was still perplexed about one thing. I knew full well from my startup days that passion alone didn't fuel this kind of highly creative and curiosity-driven research. It also required capital, a lot of it, and that didn't come out of thin air. How in the world was all of this—the professors, the students, and this unusual facility—being funded?

I got my answer when Machover introduced me to Professor Mitchel Resnick, who in addition to directing one of the lab's better-known groups, *Lifelong Kindergarten,* was also head of the Media Arts and Sciences academic program of the Media Lab. We met in his office in "the cube": a double-height, studio-like space that was used for live performances in the early days of the Media Lab but had long ago been converted to cluttered quarters for three research groups, including Resnick's.

As I walked through his workshop on the way to his office nestled in the back, I couldn't help noticing a huge array of open cabinets filled with multicolored LEGO bricks, and a collection of them were scattered on the floor and in various states of assembly. This looked a lot like the kind of hands-on play area one might see at an FAO Schwarz store during Christmas. When I commented on this, Resnick explained that the LEGO Group had been a "consortia sponsor" of the Media Lab since the lab's founding. This unique financial arrangement meant that in return for paying an annual subscription fee, LEGO was granted unlimited rights to all intellectual property (IP) developed at the Media Lab while they were members, in a highly unusual "open IP" policy that was pioneered by the lab's founders and still had yet to be duplicated anywhere. Perhaps as important, employees from these sponsor companies were entitled to collaborate in the research process itself, interacting as often as they liked with the faculty and students at the lab. In the case of LEGO, this was primarily with the Lifelong Kindergarten group—LEGO was intensely interested in the link between creative play and learning that the group explores—but also with any other group, faculty member, or student that caught their interest. LEGO was one of more than sixty such corporate sponsors from a wide variety of industries who were members of the Media Lab at that time, effectively agreeing to subsidize and share evenly in the intellectual fruits of the lab's labors.

· · ·

But now another question was bothering me. Given how much money corporate sponsors were pouring in, who was actually setting the research directions pursued in the Media Lab? Wasn't there a risk of the researchers being beholden to the short-term business interests of the companies subsidizing their efforts? Resnick assured me that the faculty and students enjoyed total creative freedom to invent and create as their curiosities and passions dictated, that sponsors like LEGO didn't *direct* the Lab's work but rather *informed* it by keeping the researchers up to speed on the needs and trends they saw in their marketplaces. This arrangement was essentially win-win. The researchers gained by having a practical context for their work, and the sponsors gained by having access to a rich set of ideas and inventions that could lead to innovative new products and maybe even make inroads into entirely new markets. As we walked out of the cube on the way to my last meeting of the day, Resnick pointed to a LEGO Mindstorms robot construction kit perched on top of a parts cabinet, and he proudly told me that the inspiration and core technology for this popular educational toy had come out of the group's long-term collaboration with LEGO.

This conversation with Resnick had my head spinning even more. I suspected that there was still a huge amount to learn about this idiosyncratic place, but by now I was convinced that there was something incredibly intriguing going on here. As far as I knew, the type of curiosity-driven, passion-fueled exploration that I'd witnessed that day had all but disappeared from both academia and industry. With more and more emphasis on short-term quarterly earnings, many big industrial research labs had been either shuttered or scaled down, and most companies had all but eliminated their research departments. Everyone in the business world was talking about innovation—it seemed there was a new book being published about it every week. But the reality was that the "seed corns" of innovation—creative new ideas and inventions—were no longer being planted as they were before. Yet many of these same companies that were pulling back on their own research internally were generously supporting the Media Lab's, and some had been doing so for the past twenty years.

As I headed home later that day, I remember feeling that excitement and anxiety in the pit of my stomach that hits when my life is about to change, when despite logic or prudence I am about to exit my comfort zone. I decided at that moment that if they were willing to consider me for the next director of the Media Lab, I was more than willing to throw my hat into the ring. Less than two months later, in early February 2006, I was back in the lower level of the Wiesner Building, this time standing in front of the lab's entire population assembled in the Bartos Auditorium, being introduced by Machover as the new director.

RUN BEFORE YOU WALK

The mess of wires, fiberglass struts, and Velcro littering the floor of Elliott and Marecki's corner of the Biomechatronics workshop in the Wiesner Building on that summer day two and a half years after I joined the Media Lab were the incarnation of the Biomechatronics group's latest project—"our latest adventure," as Professor Hugh Herr, who heads up the group, puts it.

I first met Herr during that exciting daylong visit to the Media

Lab in December 2005 I just described. Tall, with an athletic build, he speaks in such a deep, low voice that you need to lean in closely to hear him. As I was to learn later, he also has a flair for the dramatic. Our first meeting began in his fourth-floor office, but he soon offered to give me a tour of the Biomechatronics workshop, four floors below. "It will be faster if we walk," Herr said, leading me to the staircase. He took off, bounding effortlessly down the steps, as I struggled to keep up. When I finally caught up with him in the workshop, Herr was already sitting at a worktable, tinkering with a piece of hardware that resembled a mechanical joint of some kind. He immediately cued up a series of videos that showed amputees trying out the group's various robotic prostheses and pointed to a prototype of each device on the worktable in front of us as it appeared on the screen. Then, much to my surprise, Herr himself appeared on screen as one of the testers. He grinned, leaned back in his chair, and casually lifted one leg onto the table, and he drew back his trouser cuff, revealing that he was wearing the MIT PowerFoot, a robotic prosthetic foot made of metal and carbon composite materials and powered by a high-torque motor, built-in microprocessors, and environmental sensors. He then rolled up his other cuff to reveal that he was wearing another one, also attached by thin metal rods to his leg below the knee. That was when I learned that Herr is a bilateral amputee.

As a teenager, Herr had been an avid rock-climber, and he had even planned to pursue the sport professionally when he got out of high school. He had given no thought whatsoever to going to college—he was, as he put it, a shop major. But then fate intervened. When he was seventeen, he and a friend were ice climbing on Mount Washington in New Hampshire when a fierce snowstorm hit unexpectedly. Herr suffered severe frostbite, and as a result, he lost both legs below the knee. Though devastated by his injury, he still clung to the belief that somehow, someday, he would climb again. Of course, before he could climb, he would have to be able to walk, and even just walking was then more difficult than he ever could have imagined.

He tried prostheses, but the prostheses of the day were impossibly uncomfortable. So he decided to take them apart and rebuild them, and he ended up fashioning devices that worked better than those made by the so-called professionals. Even though he had been told in no uncertain terms that his climbing career was over, that didn't stop him from inventing new prostheses for walking and even some novel devices for climbing. One was a prosthesis molded directly into a spiked foot for ice climbing (a favorite activity of his), and one that he was determined he would never give up. Incredibly, just six months after the amputations, Herr was back climbing, tentatively at first but soon better and higher and faster than he had ever climbed before. (Hugh Herr's story was eloquently told by Alison Osius in her 1991 book, *Second Ascent: The Story of Hugh Herr,* which was later the subject of a National Geographic Channel special.)

"I was told by my doctor that I would never climb again," Herr says, absolutely deadpan. "He was wrong."

Herr's doctors had told him he'd be disabled forever, but in proving them wrong, Herr realized something that came to later shape the direction of his research at the Media Lab: There is no such thing as a disabled *person.* What was disabled, he concluded, was the *technology* that was supposed to be serving the community known as "disabled." So he decided to devote his life to changing that situation. And to do so, this self-proclaimed "high school shop major" went on to major in physics at college and earn a master's degree in engineering at MIT and a PhD in biophysics from Harvard. After a stint as an assistant professor at the Harvard Medical School, Herr joined the Media Lab in 2004, and in a short time his Biomechatronics group has become world renowned for its cutting-edge work in *bio-hybrids,* the science of merging of humans and machines to create prostheses that rival the real thing. Within just six years, Herr and his students have developed a robotic prosthetic knee with the full range of motion, an orthotic that can restore normal gait for stroke victims, and the PowerFoot, a prosthetic ankle so sophisticated that it gives amputees back the ability to walk at normal speeds and normal levels of effort. Soon to be brought to market by Herr's Cambridge-based startup company, iWalk, it is already being worn by Iraqi and Afghanistan veterans who lost limbs in the wars.

Today, Herr's team has set its sights on building the world's first *exoskeleton to augment human running,* a streamlined next-generation version of the walking device developed several years before in Herr's lab to help lighten soldiers' burden of carrying heavy loads for long distances, over rough terrain. The latest incarnation of an "exo"—the team's nickname for a class of exoskeletal devices that are worn outside of the body and that augment natural human function—is intended to allow everybody, regardless of his or her age or fitness level, to run farther and faster, without breaking a sweat.

Of course, this won't be easy. A major problem with conventional prosthetics is that they typically sap energy from the body, quickly fatiguing wearers. So Herr's team of young engineers has to figure out how to make the running exo so strong, so light, and so superbly designed that it not only performs as well as the muscles and tendons that normally control running but also does it in such a way that it actually makes running easier for the body. The team has already managed to accomplish this with the PowerFoot prosthesis for walking amputees, and they have made additional progress with a device that allows continuous hopping. But the requirements here are in some ways more challenging.

Grant Elliott, Andy Marecki, and other members of their group have been laboring over this version of the running exo for months. Finally, on a steamy July afternoon, they are ready to strap it on Marecki and take their pet project outdoors. Today's test run, one of many before the final prototype will be complete, is focused on making sure the device, particularly the wetsuit shorts that serve as a harness, will be both comfortable and correctly aligned with Marecki's body. It's a prelude to tomorrow's critical test, in which Andy will don a facemask that measures his metabolic rate and determines whether the running exo enables him to perform at better-than-human capacity.

Designing a harness that is lightweight, comfortable, and doesn't ride up on the wearer is a major challenge. Skin tends to be very slippery over the bone, which makes it tough to get a secure grip around the bony areas of the waist and hips. However, Hazel Briner, an MIT undergraduate intern in the Biomechatronics group majoring in biology and engineering, solved this problem by sewing a thin lining of that same nonslip fabric that lines a mouse pad into the wetsuit shorts and under the straps that attach to them. Briner is not just an excellent engineer and skillful sewer; she really knows a thing or two about harnesses, thanks to four summers spent as an aerialist touring with the Circus Smirkus Big Top, the world-renowned international youth circus.

Marecki slips the mended pair of wetsuit shorts over his jeans, wraps the black straps around his waist, and then secures two long struts to hip plates that are sewn onto the shorts. He bends down to attach the bottom of the struts to the toe clips on his bicycle shoes. The tall, lanky engineering undergrad looks even taller as he takes wide, exaggerated steps, bending and extending his legs to test the embedded knee joint, which is a mechanical clutch powered by an electromagnet. Marecki had spent much of his summer designing this clutch because the commercially available clutches were much too heavy.

"Feels good," Marecki tells Elliott approvingly.

He takes another step and bounces up and down, looking like he is preparing for liftoff. "It feels a little like pole vaulting," he says, clearly delighted.

"Let's take it for a spin," Elliott says, gesturing toward the door. As they make their way through the corridor toward the spacious lower level of the five-story atrium, Marecki walks clumsily. It's a lot easier to run in the device than to walk, he finds.

He pauses, makes a few adjustments to secure the struts at his knees and hips, and then sets off toward the opposite wall of the lower atrium, the struts protruding from his side. As he crosses, each running step sends a "clack, clack, clack" that sounds like an old manual typewriter, echoing through the atrium. Marecki is making so much noise and looks so outlandish that even the onlooking MIT students who are used to seeing odd-looking new devices being tested in the Media Lab atrium, do a double take. A few people on the upper floors lean over the balconies, peering down to see what is causing the commotion.

"We've got to do something to dampen that noise!" Elliott winces, and he scribbles a few thoughts in his notebook.

"OK. Let's take this outside and see what it can do."

Marecki leaps into action and bounds up the steps that lead to the Ames Street plaza adjacent to the lab. He proceeds to "exo-run" up the block toward Kendall Square, reverses direction, and then strides back at a fast clip past the lab's main entrance, exoskeleton clanking loudly. At that very same moment, an orange and green Old Town Trolley tour bus, carrying summer tourists, pulls up in front of the main entrance. The bus stops, and its gawking passengers, pressed against the windows, whip out their cameras and furiously snap pictures. One passenger opens a window and shouts jokingly at Elliott, "Where can I buy one of those things?" Elliott calls his bluff, shouting back "Walmart."

A few minutes later, Elliott and Marecki are back at the Biomechatronics workshop, the running exo is disassembled, and its pieces are once more scattered on the floor. The two young inventors report on the field test to the rest of the group, and before long they are already tinkering with the design in preparation for tomorrow's demo.

Each one of these steps, though small in itself, brings Herr and his team one step closer to their goal, and indeed, the wider goal of the Media Lab itself: to invent technologies that empower ordinary people from every walk of life to do extraordinary things.

FURTHER REFLECTIONS

As you will read in the pages ahead, Media Lab faculty members continually implore their students to take risks—to let their imaginations soar and take full advantage of the extraordinary creative freedom that exists at the Media Lab. The result is that scores of risky projects are being worked on every day, hundreds every year, and many thousands over the past quarter century. The only ideas that are rejected outright are those that, for one reason or another, are *not* risky enough.

With all of these risky projects going on, there is never such a thing as something considered to be a failure at the Media Lab. Prototypes that don't work out as imagined or anticipated are used as opportunities for the researchers to learn, reimagine, or head in a different direction. Even if they are mothballed on a shelf or in a display case, or archived on the network server, these prototypes don't die; more often than not they are resuscitated in the future when someone else, probably working on a very different problem, discovers a way to utilize them in his or her own latest invention.

This creative freedom, a rarity in the world of institutional research today, is possible at the Media Lab for a single reason. The corporations who sponsor the Lab are *not* looking to us for solutions to specific problems they may have, nor are they prospecting for intellectual property that they can expeditiously turn into next quarter's product introduction. Instead, they are seeking something much more important: the chance to "drink from a fire hose of imaginative ideas and inventions,"* most of which

* This is an adaptation of a classic quote attributed to Media Lab cofounder and former MIT president Jerome Wiesner: "Getting an education at MIT is like trying to take a drink from a firehose."

appear at first to be unrelated to their core businesses, but some of which eventually translate into innovations that have a truly profound impact on people, business, or society. Since this occurs regularly over time, often through the process of serendipity that you will read about in a later chapter, the Media Lab—home to thousands of risky projects—is paradoxically not really a risky proposition at all. Its resilience over the last twenty-five years across multiple generations of technology and three economic downturns is evidence of this.

CHAPTER 2

Disappearing Disciplines

In my five years as director of the Media Lab, I've received some pretty unexpected and unusual e-mails, but perhaps none more surprising than the one that announced itself in my inbox late one Saturday evening in December 2009. It was from Professor Sandy Pentland, head of our *Human Dynamics* group, and the subject line read "We win the DARPA fortieth anniversary Internet challenge contest." My first impression was that Pentland was pulling some kind of prank, since I'd heard nothing about any such contest, but out of curiosity I clicked on the attachment anyway. It was a press release from the U.S. Department of Defense dated December 5, and it began:

MIT Red Balloon Team Wins DARPA Network Challenge

The Defense Advanced Research Projects Agency (DARPA) has announced that the MIT Red Balloon Challenge Team won the

$40,000 cash prize in the DARPA Network Challenge, a competition that required participants to locate ten large, red balloons at undisclosed locations across the United States. The MIT team received the prize for being the first to identify the locations of all ten balloons.

"The Challenge has captured the imagination of people around the world, is rich with scientific intrigue, and, we hope, is part of a growing 'renaissance of wonder' throughout the nation," said DARPA director, Dr. Regina E. Dugan. "DARPA salutes the MIT team for successfully completing this complex task less than nine hours after balloon launch."

It took me several read-throughs to accept that this was probably not a hoax, and just to be sure, I replied to the e-mail asking Pentland to give me a call any time. My cell phone rang immediately, and it was indeed his voice. He explained that "the MIT Red Balloon Team" referenced in the press release was in fact his group at the Media Lab, and he said that earlier that day they had beat out 4,000 or so other teams from around the globe who had been preparing their approaches for several months. I asked him why I had never heard about this until now, and he explained: "I didn't either until Thursday, and my guys learned of the contest only on Tuesday." I hung up and shook my head in disbelief.

But as I thought about this over the weekend, it began to make much more sense, especially after I reread his e-mail in which he had described the victory as "an amazing example of what social network design and clever mathematics can do." I realized that this challenge was not really that far afield from the heart of Pentland's own research.

With an eclectic background in behavioral psychology, computer science, and mathematical modeling, Pentland has devoted his career to finding new ways to understand people and predict their behavior by combining the tools of all three disciplines in creative ways. Even so, it was clear that there must be an interesting story behind how he and his researchers pulled off a victory in the Red Balloon Challenge, and I wanted to hear it, so I arranged to meet with the team first thing Monday morning.

Riley Crane, a physicist and postdoctoral fellow in the Human Dynamics group, had spearheaded the effort and did most of the talking. This was my first time meeting Crane, and I remember thinking that he reminded me of the Jeff Goldblum character in the *Jurassic Park* film, an eccentric mathematician who, like Crane, was tall and lanky, wore dark frame glasses, and projected a sort of rakish brilliance. I asked him to take me back to the beginning of the story. The challenge had been posed on October 29, 2009, by DARPA (the Defense Advanced Research Project Agency), the research arm of the Department of Defense. It was issued to mark the fortieth anniversary of the birth of the ARPANET, the direct predecessor of today's global Internet, behind which the agency was the creative force. On that calendar date, forty years before, at 10:30 p.m., the first message ever was sent from a host computer at UCLA to one at the Stanford Research Institute over a network that consisted of only four "Interface Message Processors," the forerunners of today's routers (of which there are now millions).

The objective of the contest was simple: "To prove just how powerful the Internet had become in the forty years since that first message," according to DARPA director Regina Dugan, who announced the challenge at an anniversary celebration at UCLA. The agency was offering a $40,000 reward to the first person or team that could pinpoint the exact location of the ten, eight-foot red weather balloons that were to be inflated and moored in ten locations throughout the continental United States on the morning of December 5. Naturally, no team could be expected to find the balloons on their own; the whole point of the challenge was to come up with creative ways to recruit and mobilize a far-flung network of people to work together to find the balloons in the most efficient way possible.

Crane told me that he hadn't heard about the contest until Tuesday, December 1, just four days before the balloon hunt, when he received an e-mail about it from a friend in Switzerland. By that time, thousands of people from around the world had been designing and planning their approaches for more than a month. There were teams with names like "I Spy a Red Balloon," "Army of Eyes," "Dude It's a Balloon," and "Google and Friends" from literally every type of institution you could think of: universities, industrial research labs, high schools, adventure groups, ballooning clubs, cartography societies, and many more.

Crane went on to explain that when he learned about the DARPA contest, he immediately made the connection between the balloon hunt and his research interests, which focused on how ideas spread in social systems. He was fascinated by how some people become "superspreaders" in what he calls "social epidemics" and how in these human networks many small actions can be aggregated to accomplish very big things. Organizing a massive search to find the balloons, he reasoned, was tantamount to creating a social epidemic on the Internet to spread the word to as many people as quickly as possible. Crane forwarded his friend's e-mail message to his colleagues in the Human Dynamics group along with the simple question, "Does anyone have a $40,000 idea?" Later that day, the group members assembled to brainstorm approaches. They knew the key would be to both come up with some sort of clever incentive mechanism to motivate as many people as possible to look for the balloons and to find an efficient way to organize that network in their search.

This is where the diverse backgrounds of the team came into play. Post-doc Manuel Cebrian had studied human behavior using the emerging field of network science, and PhD student Galen Pickard had degrees in electrical engineering and neuroscience. Both of them were familiar with the psychology of reward schemes. After some discussion they posed the following question:

What if we were to create an incentive scheme that would reward not just the finder of a balloon but also the chain of people leading up to a successful finding?

Pickard told me that psychologists call this a *temporary recursive incentive scheme,* and using the context of the challenge, he explained how the scheme works. The basic concept was to distribute the rewards across the network, according to a system that allocated $4,000 of the reward money for the "chain of finders" of each balloon. The first person to find a balloon and report its coordinates to the team earned half that, or $2,000. But here was the ingenious part. The *person who recruited the person who ended up finding the balloon* into the team's balloon finding network received half of that, or $1,000, and the *person who had invited that person* received $500, and so on. This meant that everyone was motivated to invite as many people as possible. It also meant that even if you were in no position to find a balloon (let's say you lived in

Europe), you could still earn money if someone in your chain found one. In short, the scheme created a financial incentive for people to attract as many friends as they could *and* convince those friends to attract as many other friends as they could. Regardless of how long the chain grew, the total payout per balloon would never exceed $4,000, and the challenge team decided that all funds remaining after the chain was paid would be donated to charity.

There was just one more piece of the puzzle: how to connect this ever-widening network of people? The team decided to set up a central website (http://balloon.mit.edu) where anyone who signed up for the balloon hunt (at no cost) would automatically get his or her own personal URL within that umbrella site (for example, http://balloon.mit .edu/frankmoss/). Participants could easily send their URL to people in their existing social network via e-mail, Facebook, Twitter, blogs, and so on, and they would then be linked not only to the person who invited them but also to everyone else in the wider network. So the central site essentially became the *viral carrier* that allowed each user to *infect* and track a growing chain of friends and friends' friends, and so on, that could lead to someone who would spot a balloon.

The group was confident in their approach, but they felt that the odds of their actually winning were slim. "We charged ourselves with the task of building an army of people within a few hours," Crane explained. "We had only thirty-six hours before the launch of the balloons to actually amass people on our side and to make them aware of us. That's a daunting task, and we thought there was no way that we could win this, but our idea was so compelling we thought that we just had to try it." On the slim chance that their mission was successful, they realized, it could offer a useful model for all kinds of other collaborative projects or social movements.

If the team was clever in their design, they proved to be even more impressive in their implementation. Once the plan was in place, Crane and master's student Wei Pan immediately began to code the website, and as soon as they had it ready for launch by late in the day Wednesday, the five original team members started firing off e-mails to their friends urging them not only to join but also to invite everyone in their address books to do the same. While Crane and Pan continued coding, Pentland asked Human Dynamics PhD student Anmol Madan, who had business experience in several startups, to begin looking into nontechnical issues such as human subject approval and tax implications. By the end of the day Wednesday, the site had about 100 registered users, and by Thursday morning the figure had reached 1,000. By Friday night, the day before the contest, the number of registered users had blossomed to 4,000, and word of the MIT Red Balloon Challenge Team had spread so widely that mentions of it began popping up all over the Internet.

By Saturday morning at 10 a.m. EST, the time of the balloon launch, there were 5,000 people registered on the website, and the number of hits was approaching 100,000. Still, given how many people they were up against, the team felt their chances of winning weren't so great. But lingering anxiety soon turned to optimism as reports came in that one balloon after another had been found. Some were spotted in likely places, such as Union Square in San Francisco and Collins Avenue in Miami. Others were spotted farther off the beaten path, including a park overlooking the Mississippi River in Memphis and a baseball field in a remote Houston suburb. Within eight hours and fifty-two minutes, Crane and the Media Lab team had successfully pinpointed all ten and had reported their exact locations to DARPA, just narrowly edging out the second-place I Spy a Red Balloon team from the Georgia Tech Research Institute (which had found nine balloons by that time).

As you probably gathered by now, the DARPA balloon challenge was not about finding balloons. It was about spreading information and using social networks to harness mass collaborative action. As Crane mused, "If one begins to contemplate the geographic and social arc traveled by information on just a single balloon's location, one really begins to appreciate the extraordinary power of these new forms of communication as well as the complex connections that bind us all together."*

That chain, which began when a friend from Switzerland e-mailed Crane a link to a *New York Times* article on December 1, ended up crisscrossing the world through e-mail, blogs, newspapers, Facebook, and Twitter before coursing through the cellular phone network and ultimately into the team's database on December 5. "If this is what can be accomplished by a few geeks and one computer in an office in Cambridge, Massachusetts, in the span of four days, imagine the kind of problems we'll be able to confront—ranging from finding missing children to real-time disaster mapping—as we learn how to harness and channel these new powers for the betterment of society." Crane added, "Win or not, we knew we had hit upon something big."

In the days following the victory, Crane did a number of media interviews, including an appearance on the *Colbert Report*. Colbert half-jokingly asked him if this strategy could be used to do anything else

---

* Crane's observations were certainly highly prescient, coming a year before the social media fueled the Egyptian revolution of 2011.

other than finding red balloons, like finding a cure for cancer. Crane initially dismissed the idea, but upon reflection, he now thinks that the technique could well play a role in cancer research. He envisions a website similar to the one used for the balloon challenge where people with all kinds of skills, fields, and areas of interest could pool their knowledge and collectively unlock some of the mysteries of cancer that still elude even the top experts.

"Maybe you can cure cancer this way, if it's done right," Crane speculates. "And if it's done on a large enough scale."

BREAKING DOWN SILOS

If you stand across the street and look at the new glass- and aluminum-wrapped Media Lab building, you might feel like you have x-ray vision. That's because you'll be able to see straight through the building, right out to the next block. The view from inside the building is just as expansive. Most interior walls are glass, the stairways are all out in the open, and the sightlines are nearly uninterrupted. Even the elevators are transparent. If you look up at them from the main atrium, you'll see not only their occupants but also the mechanical guts of the elevators themselves as they glide up and down the building's six stories. There are no confined spaces anywhere, no optical boundaries separating one workshop from the others.

The design of the building itself speaks to how we do things at the Media Lab. What I mean is that the boundaries between fields, disciplines, and areas of research are as equally nonexistent as those between workshops. And the backgrounds of the people here are as diverse as the projects they are undertaking. The lab isn't populated just by the computer scientists and engineers you'd expect to find at a place like MIT. It is also filled with people schooled in a dizzying array of fields—architects, musicians, social scientists, psychologists, designers, neuroscientists, physicians, economists, physicists, visual artists, writers, performers, and much more—many of whom are working on projects that *seem* to be completely outside their realms of expertise.

That's exactly the idea. Putting people with this diverse range of backgrounds and interests in an intellectual environment as transparent and open as the physical space we inhabit is fundamental to our approach. This was inspired at the outset by Jerome Wiesner, a cofounder of the Media Lab and an ex-president of MIT. Wiesner himself was a one-person multidisciplinary team. He was an eminent scientist (in a broad array of fields ranging from information and communication technologies to cognitive and brain sciences), a distinguished statesman (John F. Kennedy's science advisor and untiring activist for nuclear disarmament), and a humanist with a deep passion for the arts.

Sandy Pentland, head of the lab's Human Dynamics group, has coined the word *anti-disciplinary* to describe the style of research at the Media Lab. He contends that it is the only approach suited to tackling the complex challenges facing us in the twenty-first century. I couldn't agree more. Today's problems—from global poverty to climate change to the obesity epidemic—are more interconnected and intertwined than ever before, and they can't possibly be solved in the academic or research "silos" of the twentieth century. In a world where problems are so complex and multidimensional, so must be the solutions, which is why the boundaries between discrete, insular disciplines as we have known them in the past must disappear.

As Pentland explains, "In the real world of today, you can't say, 'It's more of a computer problem, so I'm going to bring in a computer scientist,' or 'It's more of an economic problem, so I'm going to bring in an economist.' You're expected to say, 'Let's look at the problem and let the problem dictate what needs to happen.' Then you go out and recruit whatever tools, knowledge, and people that you need to get the job done. That's what I mean when I use the term *anti-disciplinary*."

Pentland's own area of research is perhaps the quintessential anti-disciplinary field. Called *computational social science,* this hot new field of inquiry takes a data-driven approach to understanding and predicting human behavior. The key to understanding people, according to Pentland, is to track their actions, and that is more feasible today than ever before by following the data that they leave behind. Computers, smart phones, GPS devices, embedded microprocessors, sensors—all connected by the mobile Internet—are forming a "societal nervous system" that is generating a cloud of data about people that is growing at an exponential rate. Every time we perform a search, tweet, send an e-mail, post a blog, comment on one, use a cell phone, shop online, update our profile on a social networking site, use a credit card, or even go to the gym, we leave behind a mountain of data, a digital footprint, that provides a treasure trove of information about our lifestyles, financial activities, health habits, social interactions, and much more. Taken together, this data creates an increasingly rich mosaic of our lives, offering intriguing new insights into human behavior, both as individuals and groups. Pentland believes that these insights will spawn revolutionary solutions to both longstanding and new challenges facing society and business today.

To this end, Pentland has assembled a highly diverse set of researchers in the Human Dynamics group, which includes students with backgrounds ranging from computer science to physics to electrical engineering to economics to neuroscience. Taking a teamwork approach to their research, they have invented a new device for computational social science experimentation that they call the *sociometric badge.* About the size of a Wii controller, you would wear this compact electronic device around your neck, and the badge is embedded with motion sensors, microphones, and other sensors capable of measuring *social signals,* including who and how many people you interact with face-to-face on a daily basis, your physical proximity to other people, the length of your conversations, the intonation of your voice, and ultimately if you tend to dominate conversations or defer to others. Your badge would then communicate this data to a base station via an infrared transceiver.

Taken separately, this information doesn't seem to be particularly significant, but when aggregated, visualized, and analyzed in creative new ways, it can actually speak volumes about how our behaviors and communication styles affect our personal relationships, our status in an organization, and how well an organization is functioning on the whole. Pentland refers to this as *reality mining,* and it is the research focus of his PhD student Ben Waber, who is using it to come up with new ways to get people in companies to work together better, and at the same time get more enjoyment out of their work. In one experiment, Waber and his fellow students from Human Dynamics outfitted twenty-two employees of a German bank marketing division with sociometric badges to track their daily interactions for four weeks. At the same time researchers monitored the number of e-mails the subjects received daily, and at the end of each workday, the workers were asked to fill out simple questionnaires about their productivity and job satisfaction.

When Waber and his colleagues looked at the results, they found that many of those who received an unusually large volume of e-mail also had as much as ten times the number of face-to-face interactions as their colleagues. These employees were the *social connectors* in the organization, the people who serve as an important focal point for the flow of information. But surprisingly the questionnaires also reported that these social connectors were actually the most *dis*satisfied of all the employees. Paradoxically, they didn't feel well liked and well connected. Instead, they felt overburdened, overloaded, and extremely stressed. Armed with the data, the researchers convinced the bank manager to reorganize his teams so that these social connectors were distributed more evenly throughout the building, actually reducing the quantity of their daily social interactions but increasing the quality. They were then able to become part of a more cohesive and ultimately more supportive network of business associates and friends. Not only did their level of job satisfaction rise but in addition the entire branch enjoyed a 30 percent increase in quarterly productivity numbers.

Waber sounds more like a psychologist or management consultant when explaining why this unexpected outcome could not have been achieved through a traditional, single disciplinary approach. "A hardcore computer scientist," he explains, "would be interested in recognizing information overload, but he or she may not be particularly interested in doing anything about it. On the other hand, someone in business school may have a lot of theoretical background that dictates the way you're supposed to look at organizations, but he or she may not be skilled with working with human behavioral data. It gets really interesting when you bring everybody together to tackle the same problem because that's when we begin to think about the problem in a different way."

Thinking about the problem in a different way—that's exactly the point of the Media Lab's anti-disciplinary ethos, which liberates its researchers to take risks and attack a surprisingly wide range of problems in ways that defy the conventional wisdom. Since joining the Media Lab, I've come to believe strongly that the key to coming up with game-changing innovations lies not in finding novel solutions to known questions but, rather, *in posing novel questions.* Only by ignoring the existing artificially imposed barriers between disciplines can we "completely change the frame" of the discussion and pose questions that no one has ever thought to ask before, including—maybe even especially—the so-called domain experts.

A MATTER OF LIFE AND DEATH

On a warm June afternoon Amy Farber and Ian Eslick are leaning over a desk in the *New Media Medicine* space, their eyes riveted on Farber's laptop screen. As they discuss databases, casually tossing around terms like *de-duplication* and *hashing*, it's difficult to tell which one is the computer scientist. When the conversation turns to cell biology, it's even more difficult to tell—and to believe—that neither has any background in medicine.

They stop talking shop briefly to compare notes about their children. Farber's then two-year-old daughter is about a year younger than

Eslick's twin girls. For a moment, they are simply two parents, chatting about how well their children are sleeping through the night. Soon, though, they are back at work, eyes once again focused on the computer screen. Farber, who has a PhD in social anthropology, acknowledges that she had little experience with computers before undertaking this project, but she quickly got up to speed.

To her, there was little choice. With the steely resolve of a mother who is determined to watch her daughter reach her first milestones, she says, "I will do whatever it takes to save my life and the lives of other women with LAM."

Farber was in law school in 2005 when she was diagnosed, out of the blue, with lymph-angioleio-myomatosis ("LAM" for short), a rare and untreatable disease, most often occurring in the lungs, that strikes only women, primarily in their childbearing years. Farber had gone to the doctor for a checkup because she and her husband were contemplating starting a family. During the physical, she casually mentioned that she was concerned about upper back pain and slight upper abdominal discomfort. Her doctor did a series of routine blood tests that all came back normal. She then ordered a CT scan of Amy's kidneys, but this time the results were more ominous. The scan revealed a small mass of cysts next to one of her kidneys and a sea of small holes scattered at the base of her lungs. That's when Farber first heard the word *lymph-angioleio-myomatosis*. Still, it was only one of the many possible conditions that could produce those symptoms, so her doctors kept digging. Four more months of tests, consultations, and visits to subspecialists at the National Institutes of Health (NIH) National Heart, Lung, and Blood Institute confirmed the dreaded diagnosis. In a way, Farber was lucky. LAM is so rare it often takes years to diagnose, and often it isn't identified until it is well advanced. Many victims are never correctly diagnosed.

When Farber was told that she had LAM, the emotions one would expect to accompany the diagnosis of a rare disease washed over her—shock, fear, even anger. She was told that the cysts would eventually get bigger and destroy her lungs and that as the disease progressed, she would find breathing on her own increasingly difficult. There was, however, nothing to be done about it. There was no cure and no treatment, and there was limited ongoing research to try to find one because to the drug companies it just wasn't worth it. LAM is such a rare condition that even if the drug companies did discover a cure, the payoff would never come close to recouping their investment in the research. A lung transplant was not an option because the cysts would eventually return and destroy the new lungs. In short, the prognosis was very grim.

Farber was sent home with a bottle of multivitamins and a warning not to get pregnant because a pregnancy could accelerate the disease. The doctors gently urged her to accept the fact that her situation was hopeless.

This did not sit well with the woman who had spent much of her adult life as an international human rights activist and an advocate for AIDS patients, and she was unaccustomed to taking things lying down. She sprang into action, and she formed the LAM Treatment Alliance, an organization dedicated to fast-tracking research for a cure for the disease. It has not been an easy road. For the past five years, Farber has been battling not only her disease but also the wall of resistance erected by those who believe that a patient can make about as much of a meaningful contribution to the process of scientific discovery as a laboratory rat.

Slowly but surely, Farber and Eslick are breaking down that wall.

I was introduced to Farber by Dr. George Demetri, director of

Dana Farber's Center for Sarcoma and Bone Oncology, whom I knew from my role as a cofounder and director of Infinity Pharmaceuticals. Demetri has long been an advocate for including online patient communities in the drug discovery process. Some years before he met Farber, he'd heard a story about a patient suffering from a rare blood disease who had learned through an online posting that someone with the same disease had experienced improvement in his condition when taking Gleevec, a leukemia drug. The drug wasn't approved for his particular disease, but the patient figured he had nothing to lose, so he asked his doctor if he too could try it. When he did, he too experienced a positive response. These results were enough to motivate some studies exploring the mechanism by which Gleevec targeted a previously unknown mutation associated with the blood disorder. A clinical trial produced continued positive results, and the drug was ultimately approved to treat the disease. The clinical discovery preceded the science, which several years later identified the molecular mechanism the drug targeted. In short, an online conversation between patients about their own experiences using a drug "off label" resulted in an important medical discovery that may never have otherwise come to light.

This story had stuck in Demetri's mind: how many other undiscovered treatments are already out there? Years later, when he got a call from Farber explaining her idea for how to let patients take part in medical research, it really resonated with him, and he agreed to help. Which is why, not long after I joined the Media Lab, I got a call from Demetri about this remarkable woman who wanted to turn the drug discovery process upside down, empowering patients in ways that were never before considered possible. Did I think that technologies like social networking (just then bursting onto the scene) could help, he asked, and did the Media Lab want to be involved? The timing couldn't have been better. I had just started my New Media Medicine research group, whose purpose was to explore new ways to level the playing field between ordinary people and medical professionals. In fact, the group was founded on the very assumption that patients are the most underutilized resource within the entire health care system. I told Demitri that his idea couldn't have been a better fit for the group and, in particular, for one of my students, Ian Eslick.

Eslick, a computer scientist with a deep knowledge of the field of artificial intelligence (AI), is pursuing his fourth degree from MIT—none of them in the field of medicine. In 1996, Eslick and his office mate, both master's degree students at MIT, cofounded Silicon Spice, a startup that was acquired by Broadcom in 2000 for $1.2 billion. At first, Eslick stayed on in various capacities, but he soon realized he didn't particularly enjoy the business world or the material culture of Silicon Valley. He didn't want to use his computer skills just to make money. He wanted to use them to enable computers to work better for people. So Eslick returned to Cambridge and enrolled in the doctoral program at the Media Lab. When I met him, we immediately hit it off. We had a lot in common—similar experiences as high-tech entrepreneurs and a similar quest to find more meaning in our professional careers. He told me that he was most interested in finding a way to make computers more effective "companions" who could help experts and nonexperts alike solve complex problems. This seemed well matched with the goals of the New Media Medicine group, so he joined as my first student.

Soon thereafter, I arranged a meeting for Eslick, Farber, and me. The three of us gathered in my office on the top floor of the old Media Lab, with its window providing a prime view of MIT's iconic neoclassical great dome. The dome has been the site of many celebrated MIT student "hacks," which are essentially practical jokes with a techie twist. One morning a Cambridge police car appeared perched atop the dome, somehow lifted there in the dead of night. Being the subject of such rich campus lore, the view of the dome frequently sparked some idle banter when guests first arrived in my office. But although she was perfectly friendly, I could tell that Farber was anxious to get down to business, so we skipped the chitchat and got right to the matter at hand.

It was fascinating to hear the brilliant social scientist and the brilliant computer scientist exchange their diverse perspectives on a question that excited them both. I could see right away that their lack of medical training would work to their advantage because they wouldn't be intellectually, professionally, financially, or otherwise vested in any one idea or approach. As Eslick explains, "People who have completed ten years of training in biomedical science in medical school look at the world in a very particular way. We had no choice but to look at it in an entirely new way."

I remember Farber's explaining how every day in the life of a patient with LAM is essentially an experiment. Like the patient with the rare blood disorder who had sparked Demetri's interest in patient-driven research, women with LAM are constantly trying out off-label drugs, new types of diets, exercise regimens, and other lifestyle changes in the hopes that something, anything, might have an impact. And why wouldn't they? It's easy to see why LAM sufferers, all but abandoned by an "expert" medical community more focused on developing cash cow drugs like Propecia and Viagra, would try to take matters into their own hands. As is the case for many rare diseases, LAM patients connect often online—in chat rooms, on community websites, via social networks, and so on—to share their experiences in informal and unstructured ways. But these sites aren't just a medium for bonding and sharing support (though they are that, too). They are also a treasure trove of patient-generated information that, if used properly, might well point the way to an effective therapy or cure.

Unfortunately, most clinicians have chosen to ignore patient-generated data, paying attention to it only *after* formal studies are completed, if at all. Some clinicians have actually admitted to us that they ignore it because "we can't publish it." Plus, the limited patient-generated data that does exist is outside the reach of physicians—whether stored on home computers, locked away in personal health records, or scrawled in notebooks or legal pads—making it hard if not impossible to access. "Patients have this incredible knowledge about their bodies, but they don't have anyone asking them about their experiences or taking time to explore that aspect of their disease," Farber explains. "All this valuable information is just falling by the wayside."

It quickly became clear how much this unlikely pair had to offer each other. As a social scientist, Farber could see that the data wasn't being effectively shared and communicated, but she didn't have a solution. As a computer scientist, Eslick could address the challenge of connecting different sources of information, but he hadn't yet found a good test case for his hypothesis that nonexperts *could* use data to solve complex problems. By the end of the meeting, Eslick and Farber had put all this together and had framed the following question:

What if LAM patients were empowered to use the data emerging from the "everyday experiments" of their lives to actually develop and test hypotheses for possible therapies and even cures?

In other words, what if LAM patients could literally play the role of scientific researchers? Eslick dubbed this model of patient-driven research *Collective Discovery*. If their vision of Collective Discovery could be made a reality, then the implications for LAM patients, and indeed all patients with rare diseases, would be truly profound.

Naturally, Farber wasn't interested in waiting the years required by the conventional research process to make this a reality. Eslick was just fine with that since he was ready to roll up his sleeves and get started on his doctoral research. So they chose a practical first step that they could implement right away. In typical Media Lab fashion, they would start with a model that Eslick could design and prototype quickly, and then they would refine it as they went. Within a few months they launched the first version of LAMsight, a secure website that allows LAM patients worldwide to record and report data about their "everyday experiments" in a way that could be shared and aggregated, while still preserving their privacy. Farber and Eslick introduced LAMsight to the patient community at an international patient conference, and before long it had 160 registered patients. That might be a tiny number by the standard of today's social networks, but it was one of the largest registries of LAM patients in the world.

In time, LAMsight was enhanced to include tools that patients could use to conduct their own informal studies testing the impact that changes in their lifestyle, diet, and exercise habits have on their symptoms. This kind of "soft data," often ignored by researchers, actually offers a wealth of information about how the disease progresses and what can be done to stop or reverse it. The prospect that patients could for the first time conduct their own informal studies and collect data totally independently of the medical establishment was truly revolutionary.

Although LAMsight was still in its early field test with patients—the stage software engineers call *beta*—Eslick and Farber decided to set up a patient-driven study as soon as possible. The odds of getting the process just right on the first try were small, but this too was in the spirit of the Media Lab's philosophy of "fail fast to learn fast."

At a conference where Farber and Eslick were introducing LAMsight to the community, a group of patients and a researcher quickly identified the first test case. LAM patients have long complained that their breathing becomes more labored around the time of their menstrual cycles, but because it affects people only a few days each month, it often fails to show up on standard lung tests. A researcher in the room declared that if the patients could establish that this phenomenon was common, it would warrant further study. This was the kind of challenge Farber and Eslick were looking for, With all this data, however anecdotal, at their fingertips, they could look for patterns. They decided that for the initial study, which they aptly named "The Estrogen Study," women would use LAMsight to monitor their own symptoms and look for any correlations between difficulty breathing and their monthly cycles. As of this writing, the study, which is expected to last about a year, is under way. Even if it doesn't result in any groundbreaking discovery, Eslick feels it will be a success if it gets the attention of the medical community. "I would consider it a good outcome if a forward-thinking researcher looks at the results from this study and says, 'This is interesting,' and assigns a postdoc to look into it for six months to see if it's the basis of a clinical trial."

Insofar that his goal is also to open the eyes of the medical "experts" to the value of patient-driven research, Eslick is already succeeding. Although many clinicians and medical professionals have continued to be skeptical, a few have begun to express serious interest in the LAMsight model. This is exactly the kind of "mind shift" in the biomedical community that Farber was trying to trigger in the first place, so she and Eslick decided to strike while the iron was hot. They came up with the idea of linking the twenty-six isolated LAM registries being used by clinicians, and then to connect them to LAMsight so they could share information with each other. "What's ridiculous is that in this era of connectivity, there is no way for clinicians to share this valuable patient information," said Eslick, exasperated. "The fact that clinical information still resides in separate silos makes even these simple interactions impossible."

So Eslick and a group of twenty clinical researchers from around the world established the International LAM Registry (ILR), a companion website to LAMsight. Eslick and a small team of programmers were able to get the ILR up and running very quickly, by reusing a good deal of the software code behind LAMsight, a classic trick of software engineers. Their goal is to dramatically accelerate LAM research by enabling physicians, clinicians, and other researchers to report and share anecdotal patient information and insights and to readily identify patients appropriate for clinical trials and studies.

The combination of LAMsight and the ILR could prove to be just the thing to end Farber's urgent search for better treatments or even possibly a cure for her disease. I certainly hope so. Time will tell, but one thing that is clear from this unlikely partnership between an anthropologist and a computer scientist: that thinking outside the frame of traditional disciplines, and posing novel, unconventional questions that wouldn't occur to the professionals, or "experts" in the field, may be the best way to spur true innovation in medicine.

REINVENTING THE WHEEL, LITERALLY

If you still question whether a computer scientist like Ian Eslick has any business attempting to reinvent biomedical research, then you may be even more skeptical to learn that Ryan Chin, an architect and urban planner by training, aims to do the same for the automobile. Just as Eslick hasn't spent a day in a medical school classroom or a medical research lab, Chin, like almost all of his colleagues in the *Smart Cities* group, has absolutely no formal training or practical experience in automotive design or engineering.

My appreciation for the extraordinary power of the Media Lab's anti-disciplinary approach blossomed on a December day in 2006, less than a year after I began as director. Chin, then a PhD student, had invited me to attend a daylong Concept Car workshop being held by the Smart Cities group, which was headed by the late Professor William Mitchell, also a renowned architect and urban planner and former dean of the School of Architecture and Urban Planning at MIT. Since arriving at the lab, I had developed an affinity for the Smart Cities team because it seemed to operate a lot like the startup world to which I was accustomed. Any time I would stop by their research area with a visitor to the lab, as I often do, one of the members of the team would immediately drop what he or she was doing and explain the project inside out. When Chin notified me of the workshop, even though it was just a week in advance of the date, I cleared my schedule for the day. I was really curious to see what made Smart Cities tick.

The genesis of the Smart Cities group is a fascinating story. Mitchell had begun laying the groundwork for the group as far back as 1998 when he launched an initiative at the Media Lab called CC++, or the

Car Consortium. Consortia of this type were the standard way of organizing research at the lab, and such efforts involved picking a single audacious challenge that captured the imagination of a critical mass of students, faculty, and sponsors. This time the challenge was stated simply: Given that it had been over a century since Karl Benz introduced the first mass-produced automobile, it was time to completely rethink and reinvent the car.

At about the same time, Chin was working on his master of architecture thesis at MIT, and it was focused on what architects could learn from the automobile industry. The connection may seem a bit out there, but as Chin explains, it really isn't: "The automobile industry was already utilizing very high-tech, advanced ways of building cars with very complex shapes, and it had been doing it with great precision and accuracy. I was wondering whether we could improve the field of architecture and the building industry by adopting some of the technologies employed by the automobile industry." That's why he sought out Mitchell.

During one meeting with Mitchell, Chin mused that perhaps the exchange of ideas could work the other way around—that *automobile designers could learn a good deal from architects and urban planners* as well. After all, people who design buildings and cities don't design them in a vacuum; they understand the importance of making design functional, and suited to how people really live. This got Mitchell thinking that maybe they could help car designers understand that urban transportation is a far more complicated challenge than simply designing a cool-looking ride. The conversations between the two continued, and in 2000 Chin enrolled in the Media Lab and joined Mitchell's Car Consortium as a researcher, beginning a long and extraordinarily fruitful partnership between them.

Chin is a natural leader as well as a dutiful apprentice, and over the next three years, from 2000 to 2003, he went on the road with Mitchell to recruit industrial sponsors. Their sales pitch was simple but compelling: The best way to reinvent the car, they told potential investors and design partners, is to do so from the unique perspective of the relationship between cars and cities—in other words, by looking specifically at *how people actually use cars in cities.* If you think about it, you can see why this got people's attention. Urban congestion has become a worldwide problem that affects half the people on the planet, and the car is a major culprit. This is more than just a daily inconvenience. Traffic jams, pollution, and energy consumption contribute to a situation in which cities—where 90 percent of population growth will take place in the twenty-first century—are quickly becoming unlivable and unsustainable. The toll urban congestion is taking on our lives, our wallets, and our planet is staggering. In rapidly growing Asian cities like Beijing, where car ownership is exploding exponentially, twenty-minute commutes have become two-hour journeys. According to a study by the Imperial College of London, in many cities 40 percent of gas consumption in automobiles occurs in the search for parking spaces.

Even though their story was intriguing, several automobile manufacturers declined to participate. But Chin and Mitchell persevered and eventually managed to convince General Motors to become a sponsor. The company had a longstanding relationship with MIT, as early-twentieth-century GM president and chairman Alfred P. Sloan was an MIT alumnus. GM agreed not only to provide funds but also to provide consulting to the Smart Cities group from some of its top car engineers and designers. That turned out to be just as important to the success of the project.

With GM behind the effort, in 2003 Mitchell decided to start the Smart Cities research group. Its goal was to create a car that would not only reduce urban congestion and pollution but also be designed specifically for how people live, work, and get around in cities. This wouldn't just be a conventional car with the addition of an electric motor and batteries, like today's hybrids. This would be a completely unique creation that would overturn all of our long-held notions of what the word *car* can mean.

Remarkably, out of the team of a dozen or so students who have since joined the Smart Cities group, only one has had any formal training in automotive design. The rest have backgrounds in architecture, urban planning, mechanical engineering, computer science, electrical engineering, systems engineering, medicine, neuroscience, visual arts, business management, interface design, operations and logistics, law, ethnography, materials science, and sociology. There's not a true "car person" among them. So in the anti-disciplinary tradition of the Media Lab, Mitchell framed their goal around a provocative question that invited ideas from all these diverse backgrounds and perspectives:

> *What if you imagined the kind of city that you wanted to live in and then designed a car for this ideal place?*

How did this unlikely group of researchers from just about every field (except for the one at hand) approach this question and ultimately come up with the design of what came to be called simply the *CityCar*? It began with a series of Concept Car workshops, which the Smart Cities group has held every academic term since 2003 until today. Mitchell taught the first Concept Car workshop with renowned architect Frank

O. Gehry, a longtime friend with whom he had taught a similar workshop at MIT on urban design a few years earlier. Gehry was fascinated by the concept of designing a car especially for cities, and he had accompanied the team to many of its meetings with automobile manufacturers. He had occasion to be at MIT during this time since he was finishing construction of the Stata Center, the distinctive new home he had designed for the MIT Computer Science and Artificial Intelligence Laboratory, and he graciously co-taught the workshop on a *pro bono* basis.

These workshops were modeled after what architecture and design schools call a *charrette*, which is essentially a freewheeling brainstorming session, a method generally considered too unstructured and too "unscientific" for an academic research environment. But that's exactly why it works so well. Students were encouraged to let their imaginations run wild and think about the very essence of "the car," as it exists in the city, in a whole new way. These were no-holds-barred, anything-goes sessions of unleashed creativity. The students were operating according to only one caveat: Don't go into the design process with any preconceived notions about how a car is supposed to look. Crazy, off-the-wall, provocative questions—the kind that could come only from a group whose fields, expertise, and perspectives were as diverse as those of the people in attendance—were bounced around the room like pinballs.

It was Mitchell who posited one of the first in a long string of seemingly outrageous questions. His was about the design of the body: "Does a car have to have a sheet metal exterior?" This question set the pace for truly out-of-the-box thinking, and the students soon got into the spirit: "Could you design cars that fold up to half the size of a smart car?" "Are wheels absolutely necessary for a car?" "Does a car need a steering wheel?" "Could you design a car that interfaces to the city like a search engine interfaces to the Internet?"

For two full years, under the guidance of Mitchell and Gehry, the Smart Cities group offered up one idea after another about what the car of the future should look like, including a remotely controlled vehicle that resembled a giant tricycle with an oversized "robotic wheel" at the front. (It was not uncommon to spot this bizarre vehicle zipping riderless around the lower atrium of the original Media Lab as students put it through its test paces.) By 2005 the group was actually at a stage where it had *too many* good ideas and was having trouble deciding which to pursue. But they knew that the project still had it skeptics in many quarters and that the best way to change minds would be to build and demonstrate a full-scale, functional prototype as soon as possible. This could begin to happen only when they had locked in a single design concept. They resolved to spend the next year focusing on producing a functional design that would resolve the two most pressing issues confronting the world's cities: massive traffic congestion and choking pollution. According to Chin: "We were looking to develop the most socially responsible concept that would have the biggest global impact."

The December 2006 Smart Cities Concept Car workshop to which I was invited was a very important milestone in that process. The workshop was held in the Smart Cities research area, which was then located in "the cube," a double-height, studio-like space where live musical performances were held back in the early days of the Media Lab but that had long ago been converted to quarters for three research groups. As the attendees squeezed in among the robotic wheels and folding-car-frame models taking up a good deal of the floor space, students from the Lifelong Kindergarten group, whose LEGO-strewn area was only a few feet away, and the *Computing Culture* group, which also shared the workspace, looked on in interest. Directly behind the Smart Cities group's space was draped a huge two-story-high reproduction of Picasso's *Guernica*. Klieg lights shining down from the two-story-high ceiling illuminated the space like a black box theater. The cube looked and felt more like an artist's loft in Manhattan's SoHo than a place where a car of the future was being invented. And that was exactly the point. It is this very convergence of the visual arts, music, architecture, design, and technology that inspires many of the cutting-edge technologies that are created at the Media Lab.

When I arrived in the cube and took a seat, Mitchell and his students were already deep into conversation with Wayne Cherry, long-time head of design at GM. They were busy reviewing and critiquing the dozens of concepts the group had come up with for the design and use of the CityCar, many of which astonished and even amused me. But all were serious in the spirit of the exercise. In response to the challenge "Can you design cars that fold up to half the size of a Smart Car?" they presented many wildly imaginative designs, among them one that, with a tough, leathery exterior of overlapping shells that fold neatly into one another so that the car appeared to be contracting into itself, looked just like the armadillos I had seen wander across the road many times outside my home in Austin, Texas. To the question "Does a car have to have a sheet metal exterior" they showed a model that looked like a bumper car made out of down feathers and even another modeled on a Nike sneaker. To the query "Are wheels absolutely necessary for a car?" their conclusion was yes but not necessarily the traditional kind. Over the past few years they had literally reinvented the wheel as an "intelligent robotic wheel": one that contains the electric motor, suspension, braking, and steering apparatus—essentially all the hardware it takes to make the car run—in a sealed unit that can be snapped on and off. One model that came out of this looked like a giant tricycle, with the oversized robotic wheel at the front. "Does a car need a steering wheel?" was an easy one for a group with computer programmers in its midst. The steering wheel would be replaced by two joysticks that digitally controlled the robotic wheels. The only thing that these prototypes had in common was that none of them looked anything like a car.

Out of this bold mass of ingenuity emerged the blueprint for what was eventually to become the CityCar: a smart, electric-powered, digitally controlled, foldable, energy-efficient, two-passenger vehicle. Thanks to the self-contained robotic wheels, the CityCar has no central drivetrain it can fold to nearly half its normal length, which makes it so compact that three CityCars can fit into a traditional parking space. In the folded position, it can also be tightly stacked at designated parking areas, just like shopping carts in front of a supermarket, or luggage carts at an airport, and it can be recharged there between uses.

There are no side doors, just two side windows made of polycarbonate glass, the same sturdy material used in jet fighter cockpits and eyeglass lenses, and the oversized windshield lifts to allow passengers to easily enter and exit the car directly to and from the curb when it is in folded position. The tires are made of a new type of rubber so they don't require air and they never blow out. The four digitally controlled wheel robots each rotates 120 degrees, giving drivers all the usual maneuverability but also incredible flexibility to navigate through tight spaces and to make an *o-turn* (also known as a zero radius turn) instead of a normal u-turn, thus completely eliminating the need for parallel parking. With the CityCar, gone would be the days of idling in the middle of the street, burning fuel and blocking traffic while waiting for a fellow motorist to parallel park his or her SUV into a tiny spot.

The model that ultimately emerged also addressed the question "Can we completely change the ownership model for cars?" During the workshop, the researchers outlined their vision for a shared-use system that they referred to as "Mobility on Demand." Not unlike bicycle sharing systems, which first sprung up in cities in Europe and are now appearing in the United States, users would rent CityCars for one way transport from point A to point B on an as-needed basis.

When the Smart Cities group began promoting this concept for the CityCar, they met many skeptics. As Chin recently recalled, "When we started out with the car project, everyone said it's crazy to fold a car. Then we talked to GM about it, and they also thought it was crazy." Over and over again the group was told that it had designed a car whose time had not come, and probably wouldn't arrive for at least a hundred years. All that skepticism faded when the Smart Cities group produced a half-scale model. "No one thinks that it's a crazy idea anymore," Chin says with a grin. "And now they're saying, well, maybe it can be built in ten years, not a hundred."

Actually, the day will come far sooner than that. As you will read in the next chapter, a small-scale pilot of the CityCar will become a reality in some cities in Europe within the next few years. And I agree with Chin that within the next decade we will see an on-demand system of electric vehicles implemented in a number of large city centers of the world. If our prediction is correct, we will owe this in part to the fact that the problem was tackled by a group of people who weren't specialists in transportation or automotive design; to the fact that the young inventors of the Smart Cities group were free to think about the problem in a fresh way, unencumbered by rigid, preconceived, "expert" notions of what urban transportation means. I also believe that this anti-disciplinary approach is key to the future of transportation—one in which the information about vehicles and the people who use them is far more important than the cars themselves.

Mitchell was fond of driving this point home: Reinventions such as the CityCar could never have been developed at an automobile company, or even at a private research lab. In fact, he said, it probably couldn't have been developed anywhere other than at a place that takes such a boundary-free approach to a problem or inquiry. "One of the things that you've got to do is ask the big stupid questions that would probably get you fired if you were working for an automaker," he said, laughing. "Why does a car have to be made out of sheet metal? Why does a car have to have an engine? With the CityCar, we have eliminated both of those things. I think a starting point always is a critical analysis of the conventional wisdom and trying to figure out where you can really challenge it in a way that leads in some fundamentally new directions."

FURTHER REFLECTIONS

If you want to get a good idea of what the most game-changing disciplines of science and technology are likely to be over the next several decades, you need to go no further than within a few minutes' walk of the Kendall Square T-Stop in Cambridge. In that small area, MIT has erected five spectacular academic buildings at a cost of almost a billion dollars over the last six years.

Walk just a few hundred yards to the west, down Vassar Street, and you'll reach the Stata Center, which houses the world-famous Computer Science and Artificial Intelligence Laboratory, birthplace of many breakthroughs in *information and communication technology* that we enjoy today. Directly across Vassar sits the Brain and Cognitive Sciences Complex, the largest *neurosciences* center in the world, where Nobel Prize winners are building a comprehensive model of the human mind. As you wander back in the direction of the T, on opposite sides of Main Street you encounter the Broad Institute and the Koch Institute, the former pioneering the use of *genomics* and the latter *nanotechnology* to make breakthroughs in the quest to tame cancer.

That these centers are incubating the breakthrough discoveries of the twenty-first century may not be too surprising—you probably are already aware that information and communication technologies, neurosciences, genomics, and nanotechnology are all critical disciplines in the decades ahead. But now take a right on Ames Street, head just a few hundred yards toward the Charles River, and on your left sits the fifth of these buildings: the MIT Media Lab. As you've read in this chapter, here, in this recently completed "transparent building," many disciplines coexist under the same roof and the boundaries between them are as invisible as the walls of the building itself.

I believe it is extremely telling that ever since the new Media Lab building opened its doors, it has become the favorite destination on campus for meetings between high-level MIT administration, corporation members, and their distinguished guests. MIT has put their money, and now their mindshare, behind the Media Lab's proposition that one of *the most important disciplines in the twenty-first century will be no discipline at all.* I think that makes it a fair bet.

CHAPTER 3

Hard Fun

The crowd is watching Pranav Mistry's every move, and he is hamming it up, clearly enjoying being the center of attention. He's wearing a thin chain around his neck, but where you might expect to find a pendant there's a tiny webcam attached to a 3M pico-projector and a mirror. All of these items happen to be connected wirelessly to a Bluetooth smart phone tucked away in his jeans pocket. There are brightly colored dots painted on his fingertips: red and green on his left hand, and yellow and blue on his right.

With a wave of his right hand, a circle of colorful computer icons appears on the wall. He points to one for e-mail, and his inbox opens, right there on the wall. He waves his left hand and points to the weather icon. *Presto!* The ten-day weather forecast appears. Next, he draws an imaginary circle on his wrist with his finger, and a watch face appears precisely where a watch would be, if he were actually wearing one. Evidently it's two o'clock. He presses his thumbs and forefingers together into a square formation and frames a view, much as a movie director might, and he mimes snapping a photo. The image that he has captured appears on the wall. But there's more. For the grand finale, he flashes a big grin and shows the audience his right palm, where an image of a telephone keypad has suddenly been projected. He places a phone call to his girlfriend. The audience at Sponsor Week erupts in laughter and applause.

Mistry is demoing the *SixthSense*, his wearable, gesture-controlled device that can transform any surface—a wall, a tabletop, or even your hand—into a touch screen for computing. This invention lets you do almost anything you could do with a smart phone—make a call, take a photograph, write an e-mail or text message, update Facebook, tweet, check sports scores, anything—without ever having to actually reach for one. I remember thinking this demo felt like a scene out of a Harry Potter movie, something you'd expect to see happen in a classroom at the Hogwarts School of Witchcraft and Wizardry.

Mistry is a PhD student in the *Fluid Interfaces* group, whose focus is on seamlessly weaving technology into people's everyday lives. The group is headed by Professor Pattie Maes, who, as you read in Chapter 1, first introduced me to the atelier style of research at the Media Lab and who practices it at the highest level in the way she runs her group. Her approach begins with the philosophy that her students "are extremely curious, incredibly creative, and have tons of good ideas, but what they are missing is the good judgment that comes from practical experience in the real world." Maes sees her role in the group as being less a director and more of a mentor, whose job it is to create an environment where her students "can come up with many flaky ideas, get a chance to try some of them, maybe fail and learn something, or just maybe develop them into something that could have a real impact on people's lives."

She often begins her group meetings, held in the Fluid Interfaces workshop, with members sharing something that happened to them in the past week—perhaps a movie they enjoyed, or some stupid inconvenience that made them angry, or some cool new technology that they came across, or just any crazy idea that just came into their heads. Maes recalls the meeting in which several students were gushing about the latest commercial incarnation of the Oblong Industries *gesture interface* display technology that had been first developed at the Media Lab in the late 1990s and was featured in Steven Spielberg's 2002 movie *Minority Report*. It may have looked impressive on screen, but Maes knew that the technology required a costly, complex setup in a room specially equipped with sensors and projectors on the ceiling, and it even required learning a new language—hardly something that could be used in everyday life.

That got her thinking. What if the group could build a more portable, practical version that the average person could actually use? She envisioned something that could, for example, let her check, while shopping at the supermarket, if a product is really as "green" as the manufacturer claims—without having to reach into her pocket for her smart phone, struggle to find the right website, and wait while her browser searches for the product, all with her tired four-year-old tugging at her skirt. What if, instead, she could simply point to the package and have the information appear right in front of her, she asked the group? That's when Mistry blurted out, "What if we could just paint pixels everywhere," and somehow the crazy idea of mounting an LCD projector on a bicycle helmet emerged. To which Maes promptly responded, "Great idea, Pranav. Go build it."

Maes's response is typical of the "just build it, don't just think it" approach that is engrained at the lab, where a prototype is worth a thousand words, a proposal is actually a quick mockup, and each invention goes through a series of iterations that are constantly being tweaked and refined until we get it just right. She explains, "Maybe some of it works in the first version, and very often, some of it doesn't, but we talk about it, we critique it, and then we go off and build the next improved version as quickly as possible."

BUILDING INVENTORS

Pranav Mistry was a prolific inventor in his native India long before he stepped foot in the Media Lab. But while all the students who come to the Media Lab are highly creative and full of ideas, and they are adept at building things, many have had very little experience with the actual process of inventing anything before they arrive.

This is where "the demo" enters in, and it has been a big part of the zeitgeist of the Media Lab since its earliest days. It is fundamental to the process by which every student, regardless of his or her background, is transformed into an inventor in a remarkably short period of time. We insist that our students actually build and then demonstrate mockups of their ideas because doing so gives them a way to express themselves and also provides them an opportunity to get feedback, not just from other students or faculty or sponsors but also from any of the thousands of people who visit the lab each year. After all, if you're trying to create technologies that improve people's lives, what better way to understand if you are on target than by actually demonstrating it to people and seeing how they react? "You start with an idea, you rough out the design of the creation, and then you dive right in and actually build it," says Angela Chang, a student in the *Personal Robots* group. "Here you live your project. The demo isn't separate from what you do. It *is* what you do."

The demo is intertwined with the process of *iterative prototyping*. Here's how it works: A student or faculty member gets an idea. Instead of just talking about it, or writing it down on paper, he or she builds a rough working prototype, refining it successively as needed. Then he or she demos it, first informally, perhaps to a colleague in the researcher's group, or in a group meeting, and eventually to sponsors during our semiannual Sponsor Weeks, those intense show-and-tell affairs where the students display all their latest inventions. Of course, to some extent, every day is a demo day at the lab. Thousands of visitors—from kindergarten classes, to visiting Nobel Laureates, to representatives from some of the country's biggest corporations—pass through the lab's corridors every year, and students are expected to demo on demand. Jamie Zigelbaum, a former member of Hiroshi Ishii's *Tangible Media* group, acknowledges that students may occasionally gripe about how running all these demos can take them away from their research, but adds that "the truth is, many of us really value the ability to demo to the sponsors. It forces us to keep building new things to show, and we learn how to talk about our work with everybody."

It's not unusual for students to whip up a working demo in just a day. Tod Machover, director of our Opera of the Future group, still remembers that afternoon in 1987 when lab director Nicholas Negroponte dropped by to announce that the CEO of Apple, John Sculley, would be visiting the lab the next morning. "Nicholas said that Sculley had never been there before and that Apple wasn't a sponsor, and he wanted us to show him something cool. I said, sure." Machover and his student Joseph Chung had been toying with the idea of building an instrument that would enable a player to improvise, and be "smart" enough to fill in notes as it was being played, all under the intuitive and expressive control of the player. So they decided then and there to make it a keyboard, and they quickly drafted the initial design.

Working at a frenzied pace, Machover and Chung broke ground on the prototype, but Machover said he went home late that night to catch some sleep. "Joe worked all night," Machover recalls. "We had to do something quickly, but music is performance art. You've got to be ready to play. You've got to be perfect."

When Machover arrived at the lab the next morning, the two of them put the finishing touches on the first *hyperinstrument,* an electronic keyboard that enabled the users to spin spirals of notes from simple chords and which was radically different from anything that was available at the time. "We played it, we fine-tuned it, we played it again, and then Sculley and Nicholas walked through the door. We played this demo, and their jaws dropped. That was beginning of it all," he recalls.

BIRTH OF AN INVENTION

I think the art of invention comes pretty naturally to Pranav Mistry, whose last name literally means "carpenter" in his native Hindi and who has been building prototypes since the age of twelve, when he asked his architect father for an Atari video set. His father couldn't afford it, so instead, father and son built one together by adapting an old TV set. Ever since, Mistry's passion has been to make technology available to everyone, everywhere, from his grandmother, who still lives in his hometown of Palanpur, India, to the people in the poorest and farthest reaches of Africa and Asia who have never been taught how to use a computer. Mistry believes that technophobic elderly people like his grandmother ought to be able to send e-mails to their loved ones who, like him, have migrated across oceans, without having to struggle with a computer. He feels the same way about the poor people everywhere who have been forced to move away from their families in search of work. Which is why almost all of his inventions are aimed at enabling people to interact with information in the virtual world in the same ways they are used to interacting with objects and other people in the physical world.

This is what spurred his invention of *Quickies,* essentially smart Post-it notes on which people can write a message using a special pen that then wirelessly transmits their message through the Internet. His inventions also include the *TaPuMa,* what he calls a "tangible public map," that allows you to search for information using objects instead of words. For example, if you put a coffee cup down on the TaPuMa surface, up will pop a map directing you to the nearest Starbucks. Mistry describes this as a "Google for objects." Mistry explains: "I wanted to bring the fun of the physical world into the digital world. I wanted to be able to take the pixels out of my computer, paint them on the physical world with gestures, and then be able to play with them."

Immediately upon getting the green light from Maes to prototype what was to become SixthSense, perhaps his most important invention yet, Mistry got to work. He quickly drafted a rough design, wrote some software, and commandeered some pieces of hardware that were lying around the lab. Within a few weeks he was demoing his first prototype to Maes and his group. But as is often the case with early prototypes, this early iteration didn't even remotely resemble the sleek device that would ultimately mesmerize the crowd at Sponsor Week a few months later. This version actually looked a bit scary—that is, if you could even look straight at it. Mistry had built the LCD projector into a metal bicycle helmet, but the glare from the projector was so bright that when he wore it, Maes had to turn her head away. Not to be deterred, he built a second prototype, this time with a smaller LED projector built into a baseball cap, but it had the same problem as the first, and it also burned his head. However, these little glitches were soon ironed out. Maes remembers that it dawned on her that there must be a better way to do it. "I said, 'Pranav, why don't you decouple it from your head and try to hang the projector on something around your neck so that at least when people look at you, they don't get blinded by the light?'"

A few weeks later, Mistry returned with the third prototype. This time, the projector, mirror, and a small camera were hanging neatly around his neck, and the light from the projector was now out of the line of vision. The camera was needed because he now wanted to "grab the pixels" with gestures of his hand, and the camera determined the position of his fingertips and fed it back to a laptop. He also wore a small backpack, which held a laptop that processed the information transmitted wirelessly from the camera. Finding the backpack a little unwieldy, he decided the device needed a bit more tweaking, and by the fourth prototype, he had replaced the laptop in the backpack with a smart phone in his pocket. Now the design was finally ready to be shown outside the group. It was an instant hit. The word of Sixth-Sense spread quickly in the tech community, and for months the media was abuzz about it—so much so that Mistry was invited to present his invention at prestigious conferences across the globe, and he was inundated with inquiries from companies and venture capitalists interested in commercializing it.

The lesson from this, says Maes, is that it is impossible to do everything in the first design phase. She believes strongly in this incremental prototyping methodology, where even if you don't get it right the first time—and most of the time, you don't—you show it to people and you get their imaginations working, and they give you lots of interesting ideas for the next prototype. "Often ideas are not that interesting at first sight, but if you keep building and rebuilding them, you might just find a gem," Maes claims.

Together as master and apprentice, Maes and Mistry rise to even more extraordinary levels than they could alone. Mistry, driven by a dream to change the world and an unbridled joy for the process of inventing new things, credits Maes for reinforcing this drive in himself as well as in her other students and encouraging them to continue on what can be a long and difficult path from risky idea to working prototype to important invention. Speaking with clear admiration and respect for his mentor, he says, "She tells us that most people are afraid of new things but that they should actually be afraid of old things. She makes us believe that not taking risks is the biggest risk of all."

POWER PLAY

The Media Lab has become known over the years for its "demo or die" philosophy, but in reality, its approach to invention and innovation is far more lighthearted than this phrase implies. "Demo or die" is about turning students into productive inventors; it is not about turning them into nervous wrecks. Nor is it about pitting students against each other to see who can spin out the most projects, or create the most gadgets, or concoct the coolest inventions in the shortest amount of time. It is about embracing the virtuous cycle between playful learning and creativity. In fact, this emphasis on playful learning, one that is rooted in the vision of the Lab's early faculty, especially Professor Seymour Papert, has been central to the lab's ethos for the past quarter century.

As a young man studying in Europe in the 1950s, Papert was the protégé and star pupil of Jean Piaget, the renowned Swiss psychologist who explored the close relationship between learning and experience in childhood development. Drawing upon Piaget's work, Papert theorized that a child's learning potential is best realized when he or she is actively engaged in the playful process of building something. When Papert came to MIT, he began to apply this theory to rethinking how children should learn in the coming age of computers. People laughed when he declared, just as the very first personal computers were appearing on the scene, that children should learn how to program computers. Papert ignored the naysayers and went on to invent LOGO, the first programming language for children and the first tool to engage children in the world of computing as active participants rather than as only passive users. Professor Mitchel Resnick, a protégé of Papert and director of the *Lifelong Kindergarten* group at the Media Lab, recalls Papert describing the time he asked a child who was just starting to use LOGO, how it was going. The child turned to Papert and declared, "It's hard, it's fun." Thus, "Hard Fun" entered the Media Lab vocabulary as the term to describe a central pillar of its approach to invention—the process by which generations of students have playfully learned how to create everything from "smart" musical instruments to sociable robots, and much more.

In December 2006, Papert was in Vietnam delivering the keynote speech at a conference of educators at the Hanoi Technology University. The day after his speech, a speeding motorbike struck him as he was crossing a busy street. Although he survived the collision, he suffered catastrophic brain damage. In an ironic twist of fate, today his wife, Suzanne Massie, and many of his devoted students and colleagues are using some of the same playful, active learning techniques that Papert famously once propounded to help him regain his speech and memory functions.

Papert certainly would have gotten a kick out the workshop that PhD candidate Jay Silver and the Lifelong Kindergarten group conducted during the Spring Sponsor Week of 2009. Just picture this scene: The workshop looks like the place that you *wish* you had sent your kids to school, or wish you had gone to school yourself. Clear plastic drawers loaded with crayons, LEGO bricks, buttons, beads, paints, chalk, pipe cleaners, modeling clay, and just about anything else you'd need to make whatever cool creation you dreamed up line the walls from floor to ceiling. A powerful looking LEGO Mindstorm robot sits on a shelf next to an electronic birthday cake with candles that make music when you blow on them. A half-dozen representatives from different sponsor companies are gathered around a table. At each place sits a paper plate piled high with slices of avocado, chunks of melon and apple, and handfuls of grapes, gummy bears, and M&Ms—as well as thin pieces of wire and several dozen small alligator clips. Looking relaxed and playful in a pair of wide-legged yoga pants and sandals, Silver is standing at the head of the table, "demoing" his oddball invention: an electronic pencil called a *Drawdio* (from the words *draw* and *audio*) that lets you "draw music."

Drawdio is actually a simple musical synthesizer that uses the conductive properties of the graphite in the pencil to complete electric circuits and produce sounds. The beauty of Drawdio is that it can turn literally anything—even fruit and candy—into a musical instrument. To show how it works, Silver inserts a thin wire through a piece of cantaloupe, secures it to the Drawdio circuit with an alligator clip, and then squeezes the melon, producing, to the delight of the audience, a musical note. Soon, the visitors are demoing their own musical masterpieces. One woman has created a veritable fruit orchestra, with each type striking a slightly different key. The man seated next to her has wired melon pieces together in such a way that when they are played, they sound like an antique harpsichord. The man seated next to him has tired of the fruit and has wired a book so that its illustrations will make a tune when traced with the Drawdio.

Like many inventions at the Media Lab, Drawdio's genesis can be traced to a serendipitous series of events. In the summer of 2007, as part of a Media Lab collaboration Silver was in Bangalore, India, working at a school for impoverished children. One afternoon, on his way home from work, Silver stopped at a street market and, on a whim, bought a kit for building a *harmonium,* an electronic keyboard that is played by striking the notes with an attached metal wand. When he came home that night and began tinkering with the kit, Silver decided to saw off the keyboard side of the electric circuit to see what would happen. Much to his surprise, he discovered that he could make different sounds by touching the untethered wand to random objects in his surroundings—which he proceeded to do happily for most of the night (much to the chagrin of his roommate). By morning, he had attached two wires to the wand and had put the contraption in a plastic bag, and he had decided to bring it to school to see what the children would do with it.

The students were mesmerized by the new toy, and they began tapping the wand on every surface they could find—even each other's foreheads and arms. Then they ran outside, into the school's garden, eager to explore what a daisy or a tree or the bare ground "sounds" like. Silver was captivated by how imaginative the students were being with the toy, but he wasn't quite sure what to do with it. When he got back to the United States, a friend suggested that he try drawing with graphite on a piece of paper and then touch the drawing with the wand to complete the circuit. Silver thought this was cool and decided to attach a piece of pencil graphite to the wand; thus, the first Drawdio prototype was born.

There's no question that Drawdio is fun to use, but it's more than just fun. Because it creates a virtuous cycle of play and learning, it's a perfect example of Hard Fun. By playing with Drawdio, a child learns about electric circuits by actually building them and seeing how everyday objects can be transformed into something completely different by actually doing it. In short, it teaches children how to be creative learners. Even adults can play and learn with it. I've seen it happen. Pretty soon after he returned from India, Silver gave me one of the early Drawdio pencil prototypes. One evening, while seated at a dinner table with a half-dozen Fortune 500 CEOs, I pulled it out of my suit-coat pocket as a way of breaking the ice. By the time I got up to give my keynote speech during dessert, they had created some pretty impressive tunes on the back of their business cards, which they continued to perfect (quite distractingly, I might add) while I spoke.

Timing being everything, this zany musical accompaniment provided the perfect segue for the part of my talk about the Lifelong Kindergarten group to which Silver belongs. There is no one better than Resnick at explaining the group's underlying philosophy, which, simply put, is that kindergarten is a great model for how learning should happen throughout life. "Children in kindergarten spend a lot of time designing and creating, whether they are building towers out of blocks, or painting pictures with their fingers. So traditionally in kindergarten, children are starting to develop as creative thinkers and are also learning important ideas. When they finger paint, they learn how to put colors together. When they build towers out of blocks, they learn what makes things stand up and what makes them fall down. Creative learning isn't just a good model for learning for five-year-olds; it's a great model for children and adults of all ages."

This includes, from my experience that evening with Drawdio, Fortune 500 CEOs.

THE MOTHER OF ALL WORKSHOPS

When I take people on a tour through the Lab, I like to stop along the way and ask students to spontaneously demo their projects. I have done this many hundreds of times in the past five years, and it is one of my favorite parts of the job, especially when visitors express astonishment, as they often do, at how these students could possibly build (and continuously rebuild) these inventions—ranging from simple handheld gadgets to lifelike prosthetic limbs, to sophisticated walking and talking robots, to full-scale electric cars—with their own two hands. As my guests look around, they may see a few soldering irons, hand tools, and small power tools littering the tables, and while that may explain the small gadgets, what about the humanoid robots and the bionic ankles and the foldable cars? The students couldn't possibly build these themselves, could they, someone on the tour inevitably asks.

Well, they could, and they do, thanks to the Media Lab's communal workshop facility, perhaps the most state of the art in any research outfit in the world. The invention practiced here wouldn't be possible without it. While every group has its own workshop facilities for tinkering and simple jobs, the bigger jobs, or those requiring more sophisticated machinery, are done in this shared "digital fabrication facility" provided by Professor Neil Gershenfeld's Center for Bits and Atoms. The objects that can be built here span all the way from the microscopic (nanometer scale) to the massive (meter scale). For many years, it was tucked away in a cavernous corridor circling the lower level of the Wiesner Building. But today it spans both buildings, with much of it showcased in a glass-enclosed cube located front and center in the new building—a true testament to how central it is to everything done at the laboratory. From the second-floor balcony, only steps from the director's office, one gets a direct bird's-eye view into the center assembly space, which is big enough to accommodate prototypes as large as full-scale cars. It also has doors leading to three fabrication rooms and the loading dock. At almost any hour, day or night, students and faculty alike can be found in the communal workshop hammering, slicing, cutting, welding, grinding, blasting, and building.

As one might expect, many of our students come here with a fairly sophisticated knowledge of computer programming or hardware design, and some even have engineering backgrounds. But many have never set foot inside a real tool shop. That is why almost all students take a course their very first semester that is appropriately called "How to Make (Almost) Anything." In this hands-on course, Gershenfeld and shop manager John DiFrancesco teach new students how to use every tool imaginable, from the basic hammers, power drills, anvils, and welders found in any high school shop class to the most advanced manufacturing hardware like water jet and laser cutters, to sophisticated software like sensors, actuators, and digital controllers—in other words, everything they need to turn their ideas into reality. By the time they complete the class, even students who have never held a hammer in their hands can use just about any piece of equipment to cut, mill, bend, sand, shear, and shape just about any material into just about anything.

It's just a few weeks into the semester, and Emily, a student who has never used a power tool in her life, is in the machine shop standing over what looks like an oversized Jacuzzi but is actually a water cutter. This tool shoots out a stream of water supercharged with finely ground bits of garnet at such a high force that it can slice through six-inch-thick blocks of solid steel, wood, glass, or ceramic. The cutting is best done underwater to prevent the water and garnet dust from flying around the room, which is why Emily, who is donning waterproof protective goggles, looks like she's about to go for a quick dip. She's working under the watchful eye of DiFrancesco, who is hovering protectively. Inches away sit a sheet metal bender, a shearer, a five-axis milling machine and a three-axis ShopBot that is the size of an SUV and can handle blocks of steel, wood, and plastic as big as five by ten feet and eighteen inches thick. Because DiFrancesco and his longtime assistant Tom Lutz like to inject a bit of humor into this otherwise serious environment, on the wall next to the water jet machine is a poster of John Belushi and Dan Aykroyd in the movie *Blues Brothers,* to whom the pair bear an eerie resemblance.

A few yards away, another student is using a laser cutter to shape three cylindrical pieces of wood. These are components of a clock whose hands remain still as the numerals move. It is, he says with great pride, the first thing that he has ever built on his own. The machine he is using is strong enough to cut steel, wood, or acrylic and at the same time delicate enough to etch intricate designs on a fragile, wafer-thin surface, like a circuit board.

One of the very first lessons in the "How to Make (Almost) Anything" course is to use the water jet cutter to slice the fingers off a life-sized wooden replica of a human hand. The obvious lesson this assignment is designed to deliver is underscored by the gruesome photo of a human hand severed at the fingers that is prominently displayed in the workshop. There's no question these are powerful, potentially dangerous tools, but because building physical prototypes of ideas is so central to the Media Lab's process of invention, it's essential that each student know how to use them—safely.

Of course, not all the tools in the communal workshop look like they could have come off the set of a horror movie. For example, one afternoon Jean-Baptiste Labrune, a postdoctoral student and member of Professor Hiroshi Ishii's Tangible Media group, is showing off a small plastic camera that has just shot out of the 3-D printer. It fits around his ear like a Bluetooth headset. On a shelf near the same printer sit three chess pieces that look as if they had been hand-carved by an Old

World craftsman. In reality, they were created by scanning pictures of actual chess pieces into the 3-D printer. A second 3-D printer that spurts water-soluble wax and acrylic can produce even more complex objects, like a spring-loaded clock, which one of our students made for the sheer fun of it, or even fully operational gearboxes for cars.

A month before, Labrune tells me, he had been sitting around brainstorming with a sponsor when they had cooked up the idea of a "hands-free camera" that would enable people to share photos with each other in real time using only gestures. Instead of having to hold an actual camera, upload the picture to a phone or computer, and e-mail it to a recipient, they could merely frame their hands around the shot to send the image, via cell phone, to a website or a chat room where others could view it. In typical Media Lab fashion, as soon as Labrune came up with this idea, he brainstormed it with several colleagues, including Keywon Chung, a student in the Tangible Media group, and Amit Zoran of the Fluid Interfaces group. Labrune and Zoran sat down together at a computer, and in less than an hour they had drafted a quick design, which was sent with the click of a mouse to the 3-D printer in the communal workshop. Over the next three hours, the printer sprayed layers of resin powder and melted plastic into a 3-D model that soon took the form of Labrune's idea. Labrune, delighted with the result, waves, magician-like, in the direction of the printer and says, "At this place you go from dreaming it to building it in a matter of hours."

Peter Schmitt of the *Smart Cities* group, who spends a lot of time in the communal workshop, fantasizes about the day when, with a press of a button, a giant 3-D printer will spit out a brand-new house in move-in condition, or a day when such a printer will be used to construct space stations on distant planets. In the meantime, he "makes do" with the tools currently in the central workshop; they may not build homes on Mars, but they do take the process of iterative prototyping beyond anything I used to think was possible.

Not long after I came to the lab, Schmitt asked me to be one of two readers on his master's thesis. The project seemed extraordinarily ambitious. He had less than a year left to design, prototype, and perfect a next-generation version of the "robotic wheel" for the group's electric CityCar, which, as you'll recall from the last chapter, needed to fit all the mechanics required to drive a vehicle within the space of the rim. This included the electric drive motor, the suspension, the brakes, and the steering mechanism, as well as everything needed to operate the wheel itself—all in a tiny unit sufficiently lightweight to snap on and off a vehicle in a matter of seconds.

Interestingly, as Schmitt explains in his thesis, electric wheel hubs aren't quite as futuristic as they seem. They date all the way back to 1911, when Porsche built the first prototype of one in the carriage house of the Austrian emperor. But this and all subsequent attempts were too complex and heavy to use in a working vehicle. So there was a serious question in my mind whether this was even possible, let alone in a fraction of the time it would have taken a car company to develop it. But I should never have doubted Schmitt, who in keeping with the collaborative nature of iterative prototyping, was building on the work of his Smart Cities colleagues Patrick Kunzler, MD, and Raul-David "Retro" Poblano. They had begun by fabricating a few quick concept models on the 3-D printers. Then, looking for parts and materials that would be cheap, they had scavenged through a local junkyard, where they found a one-horsepower sewing machine motor and a spare tire and rim. Using the water jet cutter, they had cut pieces of a polycarbonate substance called *lexan* into pieces they could use to assemble those found components into a working prototype that would include an in-wheel motor and suspension and a vertical gear shaft that a third student had designed. Though this prototype demonstrated the concept, it was still rough, much too bulky, and didn't yet include steering, braking, frame connections, and control electronics.

This was a perfect challenge for the soft-spoken but indefatigable Schmitt, who "warmed up" for the project by building a bizarre-looking bike with the oversized robotic wheel in front. (In order to drive this bike, Schmitt had to lean all the way forward and grab two handles that emerged from the center of the wheel.) Several more iterations later, Schmitt came up with an ingenious design. It was lightweight, and it could be easily snapped on and off the chassis using a connector that worked like a USB plug and provided power and data communication between the robotic wheel and the vehicle's control system. Pretty soon, he had built four of these in the communal workshop, and he had snapped them onto a half-scale test vehicle prototype—which he fashioned out of laser-cut, furniture-grade plywood. By the time he turned in his thesis—amazingly, just ahead of the deadline—he had tested the functional half-scale model in action. In other words, in less than a year, Schmitt had managed to design and build something that had eluded automobile manufacturers for a century, all within the confines of the Media Lab machine shop.

FROM INVENTION TO INNOVATION

In the dead of night about a week before Christmas 2009, members of the Smart Cities group quietly pushed a car out of their second-floor workshop of the brand-new Media Lab building, into the freight elevator, and down to the lower-level atrium for its first test drive. This was the first demo of the working prototype of the CityCar, the project that had consumed the past three years of their lives.

PhD student Will Lark had the first crack at the remote control, and when he first commanded the car to move forward, it lurched so quickly that it would have crashed through one of those glass walls had one of the students not grabbed and braked it from behind. The second attempt went more smoothly, though, and soon Lark and fellow PhD student Raul-David "Retro" Poblano were passing the remote back and forth as they took turns moving the car through its paces—forward, backward, and sideways. Meanwhile undergraduates Tom Brown, Charles Guan, and Nicholas Pennycooke were helping PhD student Ryan Chin make repairs on the fly. Screws needed to be tightened; a few wires needed to be cut. The power level on the wheels needed to be adjusted so that they would have enough torque to move easily along the smooth marble floors. When a wheel started to fall apart, Chin hastily patched it together. He didn't have a choice; there were no spares.

The Smart Cities group will keep refining and test driving as many versions as it takes to get the design just right. As Chin puts it, "If a cell phone dies, no one is going to die. But there is a degree of complexity in building a car that doesn't exist with most other lab projects. There are many moving parts that have to go together. We have to make sure that the controls work, and we have to make sure that the safety systems work. Think about it. The car as we know it is the product of a hundred years of evolution, and it's a very difficult task to reinvent it with just a small team of MIT students, no matter how smart they are."

People often use the terms *invention* and *innovation* interchangeably, but they are actually quite different. *Invention* is the art of coming up with and creating revolutionary new ideas and technologies, whereas *innovation* involves figuring out how to actually execute and implement them. In other words, the robotic wheel and the CityCar are brilliant inventions, but true innovation would mean taking those inventions out of the lab, and putting them to use in the real world. Moreover, while inventions can be born from the imagination and hand of a single individual, and many are, true innovation on the scale needed to solve today's complex, interconnected, and global problems requires a larger collaborative effort among people and organizations alike. In his book *Reinventing the Automobile,*[*] Professor William Mitchell referred to these kinds of problems as "wicked problems." According to Mitchell, these are problems that "don't seem to have a clear answer, that will require consensus building, with solutions that may be in conflict with one or two key constituent groups, and will require the cooperation of large slow-moving organizations."

When they embarked on the CityCar project, Mitchell and his Smart Cities students were determined to tackle the "wicked problem" of traffic-choked, polluted urban environments by developing radically new modes of personal mobility for cities. The robotic wheel invention

---

* William J. Mitchell, Christopher E. Borronni-Bird, and Lawrence D. Burns, *Reinventing the Automobile: Personal Urban Mobility for the 21st Century,* MIT Press, Cambridge, Mass., 2010.

was the first step, but it was only the beginning. The next step was to build fully functioning working prototypes of the car and ultimately to conduct real-life pilots in cities. In other words, it wouldn't be enough to just invent the car of the future. To achieve true innovation, they would have to come up with a way to put it to use.

Just as the robotic wheel was only part of the CityCar design, the design of the vehicle itself is only a piece of the group's plan to solve the "wicked problem" of urban congestion and pollution. They must think not only about the invention of the vehicle itself but also about how it will fit into the greater context of the city and urban life—how it can be used to create an entirely new *system* of personal transportation that provides the benefits of mass transit while still preserving the freedom and flexibility of personal forms of transportation like cars, bikes, and scooters. After all, cars in cities are parked 95 percent of the time, and 85 percent of cars have only one occupant. That's a lot of empty cars.

The team's thinking was influenced by existing vehicle sharing systems, like the Zipcar network in the United States and the world's largest bike sharing system, called Vélib, in Paris, which validated that these systems could work. They wondered if they could learn from them and do it better. As a result, they are designing a shared-use system they call *Mobility-on-Demand* (MoD). With MoD, you would be able to pick up a CityCar at one location and drop it off at any of hundreds of re-charging stations strategically placed near bus, train, and subway lines. Picking up a car would be as easy as renting a luggage cart at the airport. Just swipe your credit card or ID at the charging station and drive away. A GPS tracker system would also prevent theft and vandalism, problems that have plagued Paris's bike rental system and others like it.

The group's big vision is of a synchronized, networked transportation system so well coordinated that there are always enough cars in the right place at the right time. To accomplish this, they will have to design an *urban nervous system* that could monitor the car's usage and traffic patterns and use sophisticated algorithms to predict demand at different points in the city at different times of day, while at the same time controlling supply to heavily trafficked locations, to avoid bottlenecks. For example, when a particularly packed train of commuters pulls into the station during morning rush hour or when hordes of Red Sox fans descend upon Fenway Park for a night game, the system will know how to reallocate and redistribute the cars so there's the right number of cars on hand and traffic will flow smoothly and without disruption. Such a system should also implement dynamic pricing—perhaps rentals would be more expensive on rainy days, to encourage carpooling and reduce congestion—as well as electric charging infrastructure that minimized energy consumption. In time the system could even navigate the vehicle without any effort from the driver and even integrate with other urban networks to provide fun personal services, like recommending nearby dining, shopping, and sightseeing. If widely implemented, it would totally change our concept of urban mobility, and it would be a huge step toward making the cities of the world more livable and sustainable.

The challenge of designing a MoD system like this will be to make it "smart" enough to accurately predict user demand. Solving this, too, will require a process of iterative prototyping not unlike the process that led to the full-scale model. Chin acknowledges that writing the algorithms to run this back-end part of the system is a far less glamorous task than building the physical car, but he notes that it actually may be even more valuable because it has "the clear potential to change society if done correctly." He explains, "Our system doesn't work just for our car. It is designed to be used for any one-way rental system, for any electric car or any bike or any scooter. We could be moving mules around the Grand Canyon or gondolas around Venice. We know that in the end, there will be lots of electric vehicles out there, and they're going to need a back-end system to keep them running."

Of course, the journey from the prototype to the widespread adoption in urban areas is a long one, but it is well under way. Not content to test this design only on the grounds of the Media Lab, or even on the streets of Cambridge, the Smart Cities group decided to partner with a city or cities somewhere in the world to build and actually put a fleet of CityCars into use. So in 2010 they created a partnership with a Media Lab sponsor to build twenty full-scale working prototypes of the CityCar and then deploy them in five major urban areas in Europe within three years.

Bill Mitchell was the first to concede that no one can be sure whether this particular version of the CityCar will catch on. But shortly before his death in June 2010, Mitchell told me that he was absolutely confident that his group was on the right track. "The traditional model of the automobile is exhausted at this point, and there will be radical intervention. I don't know if it will be precisely the model that we lay out, but it will be a lightweight, small-scale, and intelligent electric vehicle."

Chin and his Smart Cities colleagues seem to be in a state of perpetual motion these days—overcoming obstacles one at a time, trying to turn the CityCar from a clever invention into a profound innovation. When I asked Chin to reflect on his experiences with his mentor Bill Mitchell and the decade-long history of the project, he saw Hard

Fun—and not just hard work—in just about every step of the way. As for the enormous challenges that still lie ahead for the CityCar project, he predicted that these would be nothing less than Really Hard Fun.

FURTHER REFLECTIONS

Three of the hardest workers, and possibly the most prolific inventors, at the Media Lab during my tenure were Pranav Mistry, Peter Schmitt, and Jay Silver, all of whom you read about in this chapter. The fact that their attitudes toward work and play are remarkably similar even though they come from very different backgrounds and cultures speaks volumes about the unique environment at the Media Lab that unleashed their creativity.

Mistry, creator of the magical SixthSense gesture interface, has an almost mystical perspective on the art of invention. "For me fun is equal to the excitement of taking risks, walking on a path I haven't walked before. When I do that, it makes me happy to be alive. If we could make the distinction between work and play go away, then society would be far more productive, peaceful, and happier."

Peter Schmitt, an inventor of the robotic wheel, builds some of the most complex engineered objects I have ever seen on the Media Lab's 3-D printer, just for fun and relaxation. A few years ago he built a fully assembled clock on the printer—spit out with hands, weights, chains, gears, and pendulum ready to work—in his spare time. For Schmitt, who was trained as a visual artist at the Academy of Fine Arts in Düsseldorf in crafts such as woodworking, metal-working, and painting, fun is expanding upon his skills as a visual artist and a builder to go into territories where no one else has ever dared to venture.

Silver, inventor of Drawdio, once told me about one of his favorite books, *The Continuum Concept,* in which the author, anthropologist Jean Liedloff, describes her experience of living for two and a half years in the South American jungle with a Stone Age Indian tribe. She describes these people to be the happiest she had seen anywhere. What struck Silver was the Indians' attitude toward "work." They made no distinction between work and play, and in fact in their language, they didn't have a word for work.

When we are at our best at the Media Lab, we don't have a word for work either. We call it Hard Fun.

CHAPTER 4

Serendipity by Design

The happy sounds of baby prattle echo through the concert hall of Kresge Auditorium, an elegant, domed structure that was designed by noted architect Eero Saarinen in the 1950s and has been the venue of choice for high-profile events at MIT ever since.

> "Ga-ga ga-ga wa-wa ga-ga wa- wa-tah wah wah waaaa waa—a waaater WATER!!!!!"

"He sure nailed it, didn't he?" Professor Deb Roy, head of the *Cognitive Machines* group, is grinning ear to ear as he shows off his then twenty-one-month-old son's newly acquired mastery of the word *water*. The thirty-five-second audio clip resonating through the auditorium's sound system is a compressed recording of what Roy calls a *word birth*. The clip documents the process by which Roy's son learned to pronounce *water*, starting with the first *ga-ga* he babbled as he splashed around in the bathtub, to the eureka moment about twelve months later when he put *wa* and *tah* together. The recording elicits smiles and appreciative applause from the thousand people packed into the auditorium in the spring of 2007 to attend the Media Lab's Human 2.0 Conference, an event devoted to exploring how the coming convergence between humans and technology will transform our minds, our bodies, and even our identities.

Roy, who is standing on stage, is presenting his *Speechome* project: the elaborate audio and video home recording system he used to document his son's life from birth to age three. The name *Speechome*, of course, is a combination of the words *speech* and *home*, but it's also a nod to the Human Genome Project, the bold, decade-long program launched in 1993 to map and sequence our entire genetic code. Roy's project is, in a way, similar, albeit smaller in scale. Using his own home as a laboratory, Roy set out to map the process by which children learn to speak and, in doing so, come up with the beginnings of a definitive and comprehensive model for how we acquire the art of verbal communication.

As the dark video screen behind Roy suddenly comes to life, he points to a graph, which at first looks like nothing more than a mess of squiggly lines. In reality, however, the lines represent a whole mass of data distilled from thousands of hours of video and audio capturing the day-to-day interactions between Roy's son and the primary adults in his life—his mother; Rupel Patel, a speech pathologist and associate professor at Northeastern University who is also Roy's research partner in the study; his babysitter, Alexia Salata; and Roy himself—as they move about their suburban Boston home.

Although Roy's twenty-minute talk details the results of his ground-breaking study of how children acquire speech, some of the most interested members of the audience are not educators, child psychologists, or even parents. They are representatives from a global financial services firm, hereafter referred to as "the Bank." And what these bankers suddenly find interesting about Roy's talk has absolutely nothing to do with how children learn. Instead, they are interested in using the Speechome technology to help them better design their walk-in banking centers.

"As they were viewing the system that I was using in my home to collect data about my son, they were imagining installing a similar system in a bank branch to enable them to better understand the interaction between their employees and banking customers," Roy explains. "They pour billions of dollars into designing their retail spaces so that they can get their customers to connect better with their associates, but they have no idea if the space is working for them or not. What we're doing is shining a light on what I call the *dark matter*—that is, the vast unknown of retail space."

The funny thing is that in his talk at the conference, Roy didn't even utter a word about retail spaces, banking centers, or even the new field that Roy now calls *retail video analytics*. No one could possibly have predicted that these bankers would make any connection between Roy's research into how children learn language and the challenges faced by the banking industry. In fact, the very presence of these bankers at the conference came about in an unlikely way. A Media Lab student named Kwan Lee, a PhD candidate in Andy Lippman's *Viral Communications* group, had interned in the representatives' department at the bank the previous summer, and Lee had become friends with some of the bankers. He thought they might enjoy the Human 2.0 Conference, so he had taken the liberty of putting them on the invitation list.

Serendipitous connections like these happen all the time at the

Media Lab, and they are one of the most powerful ways in which our research evolves, but they really aren't as accidental as they seem. In fact, they are the direct result of one of the most central tenets of the Media Lab's approach to invention and innovation: what I call *serendipity by design*. Serendipity by design means that there are no such things as truly accidental discoveries; that these "accidents" happen because the Media Lab deliberately creates an environment in which unlikely connections *can't help but happen*; an environment in which the only real master plan is that there is no master plan, in which professors and their students are encouraged to constantly branch their research in new directions, follow their curiosities, cross-pollinate with others, and venture outside of their specialties. The reason these seemingly random connections between people and people, and people and ideas, happen at the lab is because when a new opportunity presents itself—for whatever reason and regardless how far afield it might be—the researchers are free to explore it and see where it takes them. Sometimes it leads to a detour, other times to a dead end. But sometimes a brand-new idea emerges, which may lead to yet another idea, and so on. The result is that any encounter or connection, by chance or otherwise, might well lead to an *aha moment* that could change the way we live, work, and play for decades to come.

A RETAIL RENAISSANCE

Fast forward to August 2009. For the past two years since that "aha moment" at the Human 2.0 Conference, representatives from the Bank have been working with Roy's group and representatives from long-time Media Lab sponsor Steelcase Corporation (a provider of high-end office furniture) to initiate a study of how customers and employees interact in banking centers, or what they call "customer-driven banking stores." To that end, a branch in Manhattan was outfitted with a version of Roy's visual tracking system—a modified version of the Speechome. Roy's system is no more intrusive than standard security cameras, which are now common in banks and ATM centers, and besides, the researchers weren't interested in tracking the behavior of individual customers. Instead, they were hoping to better understand the bigger picture of how the Bank's employees and customers interact in the physical space of the branch—just as Roy was trying to understand years' worth of interactions between his son and his son's caregivers in the physical space of his home.

Every few weeks a team of ethnographers, designers, and marketers from Steelcase would come to the New York City branch to test out various configurations of furniture in what they called the *collaborative areas* of the bank. For example, in one, employees might sit shoulder to shoulder with customers in equal-sized chairs. In another, they might sit on stools across a high desk, and so on. Then they used Roy's video technology to collect and quantify observations on how these different configurations of furniture and workspaces affected the employee-customer interactions; the goal, of course, was to figure out which one fosters the best communication and is the most conducive to closing a mutually beneficial deal.

Video recording of customers as they roam through retail settings is not new, nor is studying such recordings to design spaces that will be the most likely to trigger a sale. What *is* new, however, is the precise way the information from the video was collected, aggregated, and analyzed. Joe Branc, Steelcase's manager of workspace future technologies, likens the contrast between the old methods and Roy's to the difference between a conventional x-ray image that is read by human eyes and a high-resolution x-ray image that is analyzed by a computer. "The computer can pick up anomalies and tumors that the eye can overlook in that image, and similarly, Roy's video data analysis can reveal things that we could miss through human observation alone."

The way in which this study came about in the first place is a great example of serendipity at the Media Lab. Joe Branc was visiting the Media Lab the winter before when he ran into Professor Bill Mitchell in the hall and mentioned to him that Steelcase had just completed a rigorous thirteen-month ethnographic study of how people interact in their workspaces. Mitchell somehow knew about the Bank's interest in this very same subject, and he instantly made the connection in his mind with Roy's Speechome project. He then instigated a meeting between people from Roy's group, the Bank, and Steelcase at Steelcase University in Grand Rapids, Michigan. The three parties met again a few months later at MIT and developed a plan for the study at the customer-driven banking store in Manhattan.

Interestingly, understanding how people interact in physical retail spaces is becoming more important, rather than less, as the digital revolution progresses. As recently as just a few years ago, the prevailing assumption in the retail banking industry was that everything was going virtual. However, it has turned out that while customers are doing some banking, and most of their research, online, when it comes to making important financial decisions, they still want to come into the branch to see a physical person. This *return to physicality* isn't limited to the financial services industry, either; most of our sponsors, in all kinds of retail businesses are reporting it, including consumer electronics giant Best Buy. So the digital revolution has not reduced the number of people shopping in brick-and-mortar spaces, but it has altered how they move about and interact in those spaces. As a result, those businesses that want to stay alive in today's climate will have to take care to align their physical spaces with the changing patterns of how people relate to one another, to the space around them, and to information. Roy believes that the new field of retail video analytics will help them do this, and it may ultimately open up a whole new way of designing buildings that can adapt smartly and dynamically to the needs and wants of the people who fill them.

The customer-driven banking store study is ongoing, so as of this writing, I can't tell you whether customers prefer the worktables to be tall and freestanding or waist high and pushed up against the wall, or whether one type of furniture configuration is more conducive to closing a deal than another. But what I do know is that the unlikely and serendipitous combination of people and ideas that came together at just the right time and right place to create this project could have happened only at the Media Lab, where people are encouraged to think openly, branch their ideas into surprising new directions, and apply their inventions in new and unexpected ways.

THE LEGEND OF THE SPIRIT CHAIR

Sometimes, designed serendipity can lead to inventions that are quite literally, magic. To see how, press the rewind button back to 1967 when fourteen-year-old Tod Machover, now a world-renowned composer of "Hyper"-symphonies and robotic operas, inventor of electronic instruments, and head of the lab's *Opera of the Future* group, first heard the Beatles' album *Sergeant Pepper's Lonely Hearts Club Band.* As this teenage cello prodigy quickly realized, the multi-Grammy-winning album, with its wild special effects and many-layered sound, wasn't just a great record. It was also a technological trailblazer. Machover recalls being mesmerized by the impressive production techniques and incredible sound, but frustrated by the fact that, for the average musician, they were so out of reach; at the time, such a richness of sound could be produced only inside a state-of-the-art sound studio and definitely could not be performed live.

As it turns out, that frustration would ultimately shape the direction of the Media Lab's music program. "The Beatles were making this fantastic music in the studio. There were only these four guys, but it had fifty layers of things going on. When they performed live, they just couldn't reproduce the sound they were imagining, so they stopped performing completely," Machover recalls. "I asked, 'How can we take music that's as beautiful and complex as what the Beatles did in the studio and make it possible to create it in a live performance, so you can share it by jamming with your friends, or respond spontaneously to an audience, by shaping it with your musical intuitions?'"

This is the question that ultimately led many years later to the creation of a product you might be familiar with: the popular *Guitar Hero* and *Rock Band* video games (the latter of which recently rolled out a special Beatles edition), created by two alumni of Machover's group, Alex Rigopolos and Eran Egozy. But when Machover joined the lab in

1985, the idea that someone with no musical training could play "Eleanor Rigby" in his or her living room with nothing but a game console and a guitar-shaped controller would never have occurred to anyone. But what did occur to him was that there might be all kinds of new ways to use digital technologies to augment and enhance performers' expressive powers. So he created a music research program and began tossing around the idea of creating digital instruments that could function like virtual sound studios.

One of the earliest such instruments the group designed, in collaboration with Machover's friend and fellow cellist Yo-Yo Ma, was called a *hypercello*. It couldn't have been more different from Ma's signature 300-year-old Stradivarius: The hypercello was wired to the hilt. It had audio pickups on the inside; position sensors on the fingerboard; position and pressure sensors on the bow that responded to the cellist's touch; a wrist sensor that identified the *type* of bow stroke; and a radio transmitter on the bridge that sent data back to a host computer, where it was processed and sent to a digital music synthesizer system. But this was more than just a smart cello. It was like a symphony's worth of instruments all in one. By changing the sound and feeling of his playing, by tightening his grip on the bow, altering his stroke or shifting his position, Yo-Yo Ma could magically produce the sounds of a flute, or a human singing voice, or even fill the stage with the harmony of a complete orchestra.

The group wasn't just designing a prototype for a lab demo. This time the goal was to create a world-class instrument worthy of a world-renowned performer who Machover knew was incredibly demanding of his instruments. "If you're on stage with Yo-Yo Ma, you've got one chance for it to work, so it can't be a funky demo. It's got to be right the first time out." As Machover said, "To us it was the musical equivalent of the moon shot."

To meet this challenge, Machover teamed up with Neil Gershenfeld, then a Junior Fellow at Harvard who had recently graduated with a PhD in physics from Cornell University, and was also a bassoonist with a deep interest in future music performance. Along with Media Lab grad student Joe Chung, who had by that time developed a sophisticated software environment for building new interactive instruments, Machover and Gershenfeld developed the software and musical behavior that would allow Yo-Yo Ma to premiere Machover's new composition, *Begin Again Again . . .*, at the prestigious Tanglewood Music Festival in August 1991.

Still, unexpected challenges cropped up along the way. As a cellist himself, Machover knew that much of the expression of music in a performance depends on how the musician moves the bow, and creating the right mechanisms to track the bow's position proved to be more difficult than the team had initially anticipated. At first they equipped the bow with motion sensors, but although the design worked well on paper, when Ma actually tried to play the instrument, the thing went haywire as soon as his hand got close to the antenna, which was mounted just below the cello bridge. To solve this problem, some of the students brought the circuit back into the lab and performed an unusual experiment. In a clever simulation of a human hand, they put hamburger meat into a glove to see what would happen when it got near the antenna. It turned out that the *hand itself* was absorbing electricity. Fortunately, the amount absorbed could be measured and then be compensated for. In addition, the team discovered that an even more accurate measurement could be achieved by reversing the circuit, so that electricity was sent through the body while sensors in the air measured hand or body movement very accurately. This was a clean solution that worked superbly.

The hypercello was like a digital recording studio at your fingertips that was controlled intuitively by how the musician used the instrument. In other words, it captured not just the notes the musician was playing but the way he or she was playing them—the *expression* of the music. So when Ma played the instrument, his music actually captured the whooshing movement of the horsehair bow as it brushed against the cello strings, the shudder as each note resonated through the cello's body, and the subtle ways that Ma moved his own body in response. In sum, it caught every nuance that contributed to the richness and complexity of a virtuoso performance and translated it into digital music that could be experienced not just from the front row at the philharmonic but anywhere, by anyone.

Machover remembers the night Ma first performed on the hypercello at Tanglewood as a turning point in Machover's career. It suddenly dawned on him that the technology his group had created didn't have to be just for virtuosos like Ma or Peter Gabriel or Prince, two other artists who had been working with his group to try out the new hyperinstruments. "First, I started thinking, my gosh, we had made an incredibly sophisticated gesture measurement system for Yo-Yo and his bow, but it should be possible to actually throw the bow away and measure somebody's natural gesture. Then it hit me. I realized that given the right tools, *everybody* can create music in an instinctive way, even people who don't have musical training, and that was something that no one had ever really thought about in that way before," Machover recalls.

But it was several unintended consequences of Yo-Yo Ma's performance that is the stuff of Media Lab legend. The first occurred when famed magicians Penn and Teller, who were fans of Machover's first opera, called *VALIS*, sought out the Media Lab's help to invent a device that could be incorporated into their act. They had happened to hear about Ma's performance on the hypercello, and it gave them a wild idea. If an instrument could be rigged to sense and respond to a cellist's movements without using any visible wires or sensors, maybe a prop of some sort could be rigged with the same technology to do "magical" things in response to the performer's movements on stage. They contacted Machover, who thought it just could work.

By another stoke of fate, Joe Paradiso, who today heads the Responsive Environments group that you will soon read about, had just taken a full-time job as a researcher at the Media Lab. Paradiso was the perfect person to help accomplish this mission. Before joining the Media Lab, he was a physicist at the Draper Laboratory, a research lab closely associated with MIT that developed the onboard computer systems that guided the *Apollo* spacecraft to the moon. Paradiso is a prodigious inventor with a highly eclectic set of interests, straight out of central casting. He has designed and hand-built many electronic music synthesizers, including the world's largest modular synthesizer, which is in his living room, and electronic keyboard interface units for internationally known musicians. He composes electronic music in his spare time, and he has long been active in the avant-garde music scene as a producer of electronic music for noncommercial radio.

Machover, Paradiso, and a group of students quickly got to work on what became known as the sensor chair and more famously as the Spirit Chair. When seated in the chair, Penn and Teller announced to their audiences that they could conjure up spirits through movements of their hands and feet. It was a clever revival of the early-twentieth-century séance, in which psychic mediums would enter "spirit cabinets" where they would supposedly channel spirits—who would then make themselves known by making sounds or music.

In reality, of course, the chair was rigged with electric field sensors that could respond to Penn and Teller's movements and gestures; it was essentially a hyperinstrument you could sit in. It had a copper plate on the seat that transmitted a slight electric signal through the seated performer's body. Two poles at arm's reach were wired with electric field sensors and halogen bulbs, so if either Penn or Teller gestured with his hands and feet, the sensors would detect the motion, and, as if by magic, music—from simple melodies to a four-hundred-piece drum kit—would begin to play and the lights would flash. It was a pretty neat trick that, after a very successful debut at MIT's Kresge Auditorium, soon became part of Penn and Teller's Las Vegas stage routine, in which they performed Machover's mini-opera *Media Medium* written especially for the Spirit Chair.

The Spirit Chair would have remained just another highly sophisticated parlor trick, however, had it not been for the presence of Phillip H. Rittmueller, then an engineer from NEC Technologies Automotive Division and a sponsor of the Media Lab, at Penn and Teller's Kresge Auditorium debut. That's when the second unintended and unexpected application of the technology behind Ma's hypercello was conceived. During the show, it dawned on Rittmueller that the same wireless electric field sensing system the chair used to track the position and movement of the human body might be employed to perform a very different kind magic: making cars safer for children and small adults.

Rittmueller had long been troubled by the fact that even though the automobile airbag systems available at the time were *supposed* to be programmed to factor in the weight and size of the passengers, they didn't do a very good job differentiating between a large adult and a small adult or child, the latter two of whom could be severely injured by the force of an inflating airbag. Rittmueller wondered if a sensing system like the one used in the Spirit Chair could help an airbag system better sense when a small adult or child was seated on the passenger side, and know not to deploy. Turns out it could. So he licensed the technology, and indeed, NEC's partnership with the Media Lab resulted in new designs in both airbags and child safety seats that have been widely adopted by the car industry and have saved thousands of lives over the years.

At some point during the orientation course for all new students at the Media Lab, someone gives them a rendition of the Legend of the Spirit Chair.

THE ADVENTURES OF SENSORMAN

If there is one person in the Media Lab who you wouldn't have predicted to become involved with baseball, it is Professor Joe Paradiso, co-inventor of the Spirit Chair. He spends all his spare time reading science fiction or tinkering with music synthesizers, and he admits that he has never had the slightest interest in sports. The Media Lab's unofficial futurist in residence, Paradiso is recognized worldwide as an expert in the important field of sensor networks, and his Responsive Environments group is currently focusing on using this technology to break down the barriers between the physical and virtual worlds. One of their latest projects is to wire the Media Lab with some forty-five *Ubiquitous Sensor Portals,* as he calls them. Each portal is about the size of a thick paperback book and is equipped with cameras, audio pickups, and a smart phone–sized screen, and it is mounted flat on a wall or other surface. Once the system is fully operational, not only will anyone in the building be able to virtually "drop in" on someone at another place in the lab via the screen on the portal, but also anyone in the building, or for that matter, anywhere in the world, will be able to "drop in" on any location in a parallel *virtual* world version of the Media Lab.

Ultimately, Paradiso believes that this technology will lead us to the point where the boundaries between the real and virtual worlds have become so permeable, it will be possible to physically be in one place, but feel like you are in another. He calls this new phenomenon *Cross-Reality.* His lab is one of a handful of research groups in the world that are working to close the gap between the real and the virtual.

At first glance this may seem far afield from the world of sports. Yet a sequence of serendipitous events led to his becoming involved in America's national pastime. It all began when he was working with Penn and Teller in Las Vegas, helping them to incorporate the Spirit Chair in their act at the old Bally's—an admittedly flashier gig than his previous life as a physicist and spacecraft engineer. One day, out of the blue, the agent of the artist formerly known as Prince contacted Paradiso. He had been in the audience at the Kresge Auditorium presentation of Penn and Teller's Spirit Chair and he had subsequently shown Prince a video of the performance. Prince immediately decided he wanted Paradiso to build him one for a tour he was doing in the next month in the United Kingdom. So suddenly, Paradiso found himself on a plane from Las Vegas to Wembley, with all the components necessary to make something like a Spirit Chair packed into his suitcase. When he arrived, however, Prince's roadies presented Paradiso with a naked mannequin and asked him to wire it such that it would play music when Prince moved around on stage, such as when he gyrated his pelvis or threw a fist into the air. This wasn't exactly what Paradiso had had in mind, but he obliged, although to this day he has no idea if Prince actually used the wired mannequin in his show.

The journey continued when the technology behind the Spirit Chair soon became part of the Media Lab production of *The Brain Opera,* an interactive music experience produced by Tod Machover, with Paradiso as the technical lead. When Paradiso and Machover were on tour with the Opera in Tokyo, they were invited to a private showing of Yamaha's new *bodysuit interface,* a wearable music controller that had pressure-sensitive sensors embedded in the toe and heel of the shoe so that dancers could map their taps into electronic music. This got Paradiso thinking: Feet could express so much more than just tap. What if he could put even more sensors into a dancer's shoe and connect it wirelessly to a music synthesizer so that dancers could create evocative music with their feet just as Yo-Yo Ma did with his bow, and Penn and Teller with their hands? With that, the *Expressive Footwear project* was born. Paradiso and his students designed a complex but miniaturized sensor board that could mount on the side of a Nike sneaker. It made an impressive stage debut at a conference in Tokyo a year later, but not until after several bugs were ironed out. At one point, during a dress rehearsal, a Media Lab student who was also a dancing gymnast jumped with such energy that the batteries flew off the shoe and landed on a Japanese dignitary in the audience.

When Paradiso later showed the shoe at conferences, people began to laugh when he went down the list of sensors on the tiny sneaker-mounted board: two sets of accelerometers (low and high G), a gyro, a three-axis magnetometer, a sonar, a bend sensor, several pressure sensors, and a capacitive sensor to measure height. But one group of listeners that didn't laugh was physicians, especially those specializing in biomechanics. Paradiso took note of this, but the full range of possibilities for this intricate sensor board really hit home for him when he attended a talk at the Massachusetts General Hospital (MGH) in Boston on the subject of biomechanical monitoring of NASA astronauts. The Q&A evolved into an intense discussion between him and another member of the audience, Dr. David Krebs, a professor at Harvard Medical School and head of the MGH Gait Analysis Lab. Krebs was studying the mechanical and neural constraints of the human gait in order to develop better rehabilitation therapies for a variety of orthopedic and neurological motor problems. This sparked a series of conversations, which ultimately resulted in Paradiso's PhD student Stacey Morris's turning the dance shoe into a tool they called the *GaitShoe*, which could measure and analyze people's step and gait over long periods of time, in natural settings, at very low cost. It was one of the first models of the wearable, wireless gait analysis devices that have since become standard in the field.

What started as a musical instrument had became a performance prop and then another performance prop and then a tool for cutting-edge medical research. We are finally getting close to baseball as promised. As the gait analysis research was nearing completion, Paradiso's group was visited by a group of doctors from the MGH Sports Medicine Department who tended to local professional teams, including the

Boston Red Sox. Their MGH colleagues who had worked with Paradiso on the GaitShoe had suggested that he might be able to help them solve a critical problem. In recent years, there had been a huge increase in shoulder and elbow injuries among pitchers, who routinely throw fastballs with release speeds approaching a hundred miles per hour. Understandably, the sports doctors wanted to know what the players were doing wrong that was causing these injuries. The stakes are incredibly high. If just one pitcher in the starting rotation ends up on the disabled list for a long stretch, the chances for another World Series championship ring are greatly compromised. But normal cameras couldn't possibly capture the pitcher's precise shoulder rotations and wrist movements, at such speeds.

Paradiso was game. He put his master's student Ryan Aylward, who had been developing a wristwatch-size wireless sensor package for interactive dance, to work immediately on designing and prototyping a whole new generation of wearable, wireless sensors that could track the shoulder and wrist motions at those incredible speeds. Aylward worked quickly, in Media Lab apprentice style. He arrived at the Red Sox 2007 spring training camp in Fort Myers, Florida, with six devices ready to be attached to pitchers' arms, wrists, hands, and torsos, and he spent a week taking measurements. The research is ongoing, now being conducted by Paradiso's student Michael Lapinski, but so far the data has done a great deal to help the pitchers differentiate between "good" and "bad" motions, thereby avoiding the injuries and keeping these talented (and expensive) athletes where they belong—on the mound.

FROM GROUNDED LEARNING
TO GROUND BALLS

Still, I think there is no better example of how one idea can lead to another, and another, and another, than the work of Professor Deb Roy and his students in the Cognitive Machines group. Over fifteen years their research has branched from creating smart robots, to exploring how children acquire language, to designing customer-friendly banks, to creating tools for fantasy baseball, to developing new methods for the early detection of autism, to helping brand managers analyze audience responses to their TV ads.

How, exactly, did this happen? Roy doesn't consider himself to be a "specialist" in any of these areas. In fact, when he got involved in the latter two areas, he knew next to nothing about baseball or marketing. He acknowledges that one of the things that attracted him to the Media Lab was that he would not be forced to pick any one discipline: "Are you a designer? Are you an engineer? Are you a scientist? Well, I've been very comfortable with ambiguity," he explains.

At a very young age, Roy began mixing knowledge and theories from different disciplines when he was dabbling in designing and building robots, which he now realizes was his way of learning about both humans and machines. When he enrolled at MIT as a PhD student in the mid-1990s, he'd already decided he wanted to create a new type of robot that could actually converse with people in meaningful ways. He wanted the robot to be able to pick up language skills as a human child naturally would, by interacting with people and objects in its own surroundings, and learning the meaning of words based on their context. Roy believed—and still does—that building these kinds of machines for what he calls *grounded learning* will not only make possible robots that can truly live among and help people but also help unlock important mysteries of how the human brain itself operates. He points out that building machines to learn about living things is not a new idea. The mystery of how birds can seemingly defy gravity, for example, was unraveled only once human beings began to build flying machines of their own.

Speaking of birds, the very first robot Roy built at the Media Lab was named *Toco the Toucan*. It was a simple affair as robots go, with a cartoonish birdlike head containing a video camera for vision and a microphone for hearing, topped with a brilliant fuzzy red feather, and sitting atop a mechanical arm for a body. He looked, as I like to tease Roy, kind of like a feather duster constructed out of an Erector Set. Roy explains good-naturedly: "I was betting that if you could get the robot to actually connect with a person in terms of the meaningful use of words, that could actually transcend how the thing looks."

When Toco spoke, his beak would move, and when he spotted something interesting, his eyelids would lift. Though admittedly less sophisticated than subsequent generations of robots, Toco was programmed to act less like a bird and more like a child. How better to design an intelligent learning machine than to model it after nature's most intelligent learning machine: the human infant. Like a child, Toco learned words not just by mimicking people but by actually interacting with them and with his environment. When Roy showed Toco a red ball and said, "Look at the red ball," Toco wouldn't just repeat the words. He actually learned what "red ball" meant. And, as Roy repeatedly talked to him and showed him things, Toco, like a child, got better and better at sorting out what those things might mean. Over time

Toco built up a lexicon—what we might call a *vocabulary* in a child—in his head. Eventually, when asked to do so, he was even able to pick out the red ball from among other objects on the table.

When Toco's ability to pick up new words in this way began to level off, much as a child's does at a certain age, Roy took things to the next level. He played Toco real-life recordings of mothers talking to their small children, and he showed Toco the specific objects that were being discussed in the audio clips. The result was a brief verbal growth spurt. However, as time went on, it became increasingly clear that there were limits to what Toco was able to pick up, and Roy was beginning to feel stymied by the robot's basic inability to be, well, more human in its thinking and responses.

"I asked myself, why would anyone want to interact socially with a robot that doesn't do much?" Roy says. "That's when my group shifted our focus from robots interacting with people to observing and creating models of how *people* interact with people." Branching the Cognitive Machines group in this new direction was a dramatic move since they had developed a strong reputation working with robots that could have comfortably driven their research for years to come.

He decided to go full speed ahead. He was determined to develop the first comprehensive model of how children acquire language, starting with their very first words. Of course, he knew there were prodigious challenges. For obvious reasons, when psychologists study language acquisition in children, they typically do it in a lab setting, where they watch and record the interactions between parents and children through cameras and one-way mirrors. But clearly, children don't learn language just in the lab or even a classroom. Most language development takes place in the private moments at home, through a child's interactions with his or her parents or caregivers or through his or her curious exploration of the immediate environment. Such interactions can spark a child's imagination and trigger a developmental leap. Roy knew that to *truly* understand how language emerges, he would need to monitor a child in every waking moment.

Opportunity came knocking, this time in a truly serendipitous fashion, in 2004 when Roy and his wife learned that they were expecting their first child. What better way to learn about language development, they realized, than to conduct a study to see how a child—their child—acquired speech in the most natural of settings—their own home. Unlike conventional childhood learning studies that had contrived starts and stops, theirs would attempt to monitor *every* precious moment in those important developmental years of their child's life.

By the time their child was born, Roy had literally transformed their yellow colonial home in the suburbs of Boston into a living laboratory. He had embedded eleven omnidirectional, 1-megapixel color video cameras into the ceilings of all the rooms, and he had strategically scattered fourteen CD-quality microphones throughout the house. Although it may have *looked* like any other comfortable suburban dwelling, behind the walls were 3,000 feet of cable connecting the microphones and cameras to a 5-terabyte storage center in his basement, where the massive amount of video and audio recordings would be compiled, time stamped, and compressed into raw data. Roy's feat was impressive to me as a technologist, but even more so as a father who had barely managed to equip my newborn daughter's nursery with a crib by the time she and my wife arrived home from the hospital!

Using a combination of audio and video, the couple ultimately accumulated what Roy calls "the world's largest home video library." They were not only able to capture nearly every murmur their son made in the first three years of his life but they were also able to actually see the *context* in which each new sound was uttered—in other words, when and where their son said a word or partial word, to whom he said it, and what he or others around him were doing at the time. In case you were concerned, every room was equipped with an "oops" button that allowed either parent to turn off the cameras and microphones when they needed to.*

But the *real* challenge came after the cameras and the microphones were turned off. Now Roy was faced with storing, accessing, and making sense of the 230,000 hours of audio-video recordings collected over the first thirty-six months of his son's life. Given how unprecedented this undertaking was, no existing technologies could do the job. The group would have to invent them. First, they would need to build a data storage and retrieval infrastructure the size of those used at the time by Google and Amazon (roughly a petabyte, or 1 million gigabytes) but at a tiny fraction of the cost and complexity. Moreover, Roy's lean academic budget dictated that it would have to be maintained by a single person rather than a team of professional storage administrators. Luckily, this kind of thing was down my alley from my high-tech days, so I was able to help to recruit a consortium of storage and networking vendors to help design a new kind of *storage area network* and donate all the equipment to build it.

---

* Roy and his wife's concern for the privacy of their son led them to develop a new model by which their son "owns" all of his Speechome data, meaning that they will not release any data for outside access until their son reaches the age of consent and if he then agrees. This approach to privacy protection for this type of data has been adopted by MIT and several other universities participating in NIH funded extensions of Speechome research.

Next, the group needed to create tools that would allow just a few researchers to analyze and draw conclusions from massive amounts of video and audio data—a volume that previously required armies of experts to pour over for years on end—in a period of just a few months. So Roy's students Rony Kubat, Philip DeCamp, and Brandon Roy developed *TotalRecall,* an audio-video data browser and annotation system that could index the audio (both the baby babble and adult talk) with the corresponding video. This enabled them to match, for example, the first time Baby Roy tried to say *water* with what he and others who were near him were doing at the time—perhaps running a bath, or pouring a drink, and so on. But there was one more challenge: the team needed to find a way to transcribe all the speech contained in the massive volume of multi-track audio recordings into text format. Off-the-shelf automated voice transcription approaches were not suited for this type of audio and the cost of manual transcription would be totally out of the question. So Roy and Brandon Roy quickly developed a program called *Blitz Scribe,* which reduced large tracks of audio to "sound bytes" and used a combination of human and machine effort to achieve speech transcription rates 500 percent faster than any known technique. With these innovative tools a few researchers were able to annotate the mammoth amounts of data in a relatively short period of time.

Not surprisingly, the Speechome project has received a huge amount of media attention, but given its small sample size—one—it's perhaps not so surprising that it has also had its share of skeptics. Nonetheless, Deb Roy has held strongly to his belief that observing a child and his caregivers on an uninterrupted, daily basis revealed insights about early childhood development that couldn't have been detected before. As it turned out, he was right. For one thing, previous studies had largely overlooked how caregivers are constantly adjusting how they respond to a child as he or she becomes more verbal. Roy's study of the Speechome data, conducted with his students Brandon Roy and Michael Frank, found that after Baby Roy learned a new word, all three caregivers (both the parents and the nanny) would naturally and automatically incorporate that word into the vocabulary they used with him. Roy calls this *fine-tuning behavior,* and he believes it is absent from mainstream theory, which holds that children pick up language from their experiences in a static or unchanging environment. According to Roy, the Speechome data reveals the absurdity of that assumption. The main environment of the child is a social one, and the closer we're able to dial in and look at it with finer and finer granularity, down to the word-by-word level, we're seeing the environment shifting along with the child.

Whether or not the Speechome project will have a huge impact on the field of child language development remains to be seen. What is clear is that the technology infrastructure and tools invented for Speechome are already proving to have a number of unexpected uses, beginning with the analysis and design of retail banking branches that I described earlier.

Another use came to light when Matthew Goodwin sat in the audience of the Human 2.0 Conference while Roy was describing the Speechome. As Goodwin listened, his mind began to race. At the time he was involved with the Groden Center, a care center for children with autism and other developmental disabilities, and Roy's talk got him thinking: If he could use Roy's audio and video recording system to observe autistic children in a totally natural setting, say, while they played at home or with other children in special-needs clinics, it might open up a whole new window into their mental world. Recording and analyzing children's daily behavior in this way might even help clinicians tailor and fine-tune treatments, therapies, and interventions to each individual child.

So he approached Roy after the conference, and convinced him to visit the Groden Center with some of his students. They were immediately excited by the project, but to make it practical, they knew they would have to find a way to dramatically reduce the complexity and cost of the system Roy had built into the ceilings and walls of his home. This was more or less like shrinking a 1970s mainframe computer to the size of a laptop. But it wasn't long before the first full-scale prototype of a *Portable Speechome Recorder* was being demoed in the Cognitive Machines workshop. Its elegant design resembles a tall arched floor lamp, only instead of a light fixture, the head contains a video camera with a wide enough angle to capture a top-down view of an entire room in which a child is playing. Cameras are built into the recorder's shaft, and microphones are hidden in both the base and the head. Most important, the device is easy to install and fits unobtrusively into living or play spaces of almost any home or clinical setting.

Soon, they will launch an exploratory pilot, funded by the National Institutes of Health (NIH). First, the device will first be installed in Goodwin's home for preliminary quality control, and if all goes well, it will be placed in a half-dozen homes of autistic children. The next step will be to place Portable Speechome Recorders into thousands of homes and day care settings of both autistic and nonautistic children, of various ages and stages of development, across the world. The hope is that the resulting data, when analyzed with new software tools Roy's students are developing, will enable researchers to isolate the behavioral patterns that characterize or even predict autism.

This could be a real breakthrough for parents with children at risk for having the disorder. Today there is no blood test or body scan that can detect autism. Whereas it is believed that the behavioral manifestations of autism begin to crop up during the first twelve to twenty-four months of life, most children aren't diagnosed until thirty-four to sixty-one months, on average. This is critical lost time in which interventions could make a real difference. If the early pilots go well, perhaps a future version of the Portable Speechome Recorder could be put in the home of *every* family with a child at risk, and that will change all this forever.

The surprising applications of Roy's work don't end here, however, thanks to Michael Fleischman, who joined Roy's lab in 2003 as a PhD student. Like Roy, Fleischman was interested in building machines that could learn the meaning of words from the context in which they're used, the way human babies do. Unlike Roy, he found working with robots to be enormously frustrating. "They're constantly breaking down," he still complains, grimacing as he recalls the story of a fellow student who was working with a robotic mechanical arm named *Ripley*—a much bigger descendant of Toco—when the thing suddenly and without warning swung around and pinned the student against the wall. The emergency kill switch just beyond his grasp, the student was forced to wait for someone to come and release him.

That decided it for Fleischman. He quickly abandoned robots for something less physically threatening: video games. At first, he thought he'd use video games to study the process of language acquisition because games had had a long history in the study of semantics, perhaps most famously used by philosopher Ludwig Wittgenstein to address the meaning of words. Fleischman's idea was to design a video game that existed in a virtual world, one in which the computer could "watch" dozens of people playing the game, while simultaneously "listening" to what they said about the game. By correlating the two, the program could possibly learn the meaning of the words the players used.

Right about the time he got sick of that project, Roy came into his office and told him that he was about to videotape the first three years of his son's life. The first thing Fleischman thought was, "This guy is crazy," but his second thought was, "I want in." He started working on an algorithm that could recognize patterns of activity in Roy's home from snippets of video taken before Roy's son was born. Just as the algorithm was getting increasingly more accurate at recognizing events, Roy sat him down and told him that it wouldn't work. Baby Roy was a newborn, and it would be too many years until sufficient video data was available in the form Fleischman would need to complete a PhD thesis.

Fleischman was getting frustrated, but fortunately it wasn't long before he was inspired to take on an entirely new kind of project that came from a surprising and unexpected direction. Fleischman came from a long line of baseball fans, and he had spent much of his youth obsessed with baseball cards. Though he'd lost interest in baseball sometime around college, his new wife, Allison, was a huge sports fan, and they would often watch Red Sox games late into the night. One evening while they were watching a Sox game, he made a striking connection. He realized that broadcasts of baseball games actually have something in common with human babies first learning language. Think about how people talk to very young children. They almost always talk about what is happening here and now: "We're going to the playground," or "It's time for a nap." Well, the same is largely true of play-by-play commentary during baseball games: "Ortiz hits a towering fly ball," or "Damon makes a sensational catch." What's more, thanks to cable sports channels, baseball is on television all the time, so there would be lots of data. And it even gives the commentary in closed-caption form, so he wouldn't have to rely on speech recognition, which can be distorted by background noise.

That's when it occurred to Fleischman that all this information could perhaps be cataloged and analyzed in much the same way as Roy was cataloging his son's early attempts at language. He barged excitedly into Roy's office to tell him of his great new idea. Roy's response: "It will never work." Humans often can't tell the difference between a foul ball and a home run, so how could a machine? But Fleischman persisted. Building on the algorithm he had already designed for his aborted video games project, he designed a program that could "learn" to recognize certain events on a reel of baseball video in much the same way that children learn the meanings of words: by their context. He used easy-to-detect cues (for example, shots of the outfield and the sounds of crowds' cheering) to signal more complex patterns of action (for example, home runs), and then he linked those patterns to the words that were used by the announcers. Ultimately, the program would learn to recognize when a batter struck out or a runner stole a base, and so on.

There were complications, though, as Roy had predicted. Every time an announcer says the words *home run*, it doesn't necessarily mean that a player has just hit a home run; the announcer could be referring to a home run from the last inning, or he or she could be predicting what might happen with the next batter. (This was the same problem that Fleischman had encountered when trying to figure out whether Roy's or his wife's words referred to present, past, or future actions.) But soon he realized it was just a matter of giving the machine enough practice: If the machine could look over hundreds and hundreds of hours of video of people hitting home runs, it would eventually start to learn what an actual home run looks like. So Fleischman created software that could "watch" massive amounts of baseball video and mine it not just for home runs but also for strikes, foul balls, and any number of other events—on the field and in the stands—that make up the "language" of a baseball game. To test the system, Fleischman built a video search engine that used the link between words and video to retrieve clips based on user queries (for example, "Show me home runs"). He compared the system to state-of-the-art video search algorithms and found that his approach outperformed them all. He had his PhD thesis in the bag.

Then serendipity once again took hold. A short article about Fleischman's PhD research in the *MIT Technology Review* attracted the attention of a National Science Foundation (NSF) project manager who was in charge of a program that helped bring technology out of academia by providing seed funds for startups. He had visited the Media Lab on several occasions, and he gave Roy a call, suggesting that he and Fleischman apply for one of the NSF's Small Business Innovation Research (SBIR) grants. They decided to go for it, and six months later they were awarded the grant to start a company to build commercial products based on the technology behind Fleischman's baseball search engine. All they needed to do was choose a name for their new startup. Then one night they went out for sushi, vowing that by the time they left the restaurant, they would have made a decision. Hence Bluefin Labs was born (full disclosure: I am a co-founder and director), and believe it or not, the company uses the same technology that Roy used to discover how his son learned his first word to enable corporate brand managers to gain real-time insight into how audiences are reacting in the social media to their mass media advertising. For them, that's the last word in market research.

FURTHER REFLECTIONS

Research programs, whether in university, government, or industrial labs, commonly fall into one of two categories: basic research or applied research. However, just like its researchers, the Media Lab's approach to research defies this type of categorization. Actually, it incorporates characteristics of both while adding its own special sauce, and that is what makes it such a powerful engine for innovation in today's world.

On the one hand, like basic research, it is driven almost entirely by the passion and curiosity of its researchers, whose goal is to increase the understanding of fundamental principles without the pressure to come up with immediate commercial applications. Also like basic research, it relies heavily on the process of serendipity to help ideas and discoveries find their way from the laboratory into practical use in the real world.

On the other hand, it has several things in common with applied research. It is informed by the needs and requirements of its corporate sponsors (although they do *not* direct it per se) and researchers build and iterate working prototypes that they test with real people in the real world. Also like applied research, the Media Lab's approach is multi-disciplinary, while most basic research is focused on a single discipline.

The unique combination, as it is applied in the Lab's open and transparent environment, is a significant "innovation in innovation." One way to look at it is from the perspective of the Media Lab's corporate sponsors. They are not partnering with the Lab to get solutions to specific problems handed to them on a silver platter, but rather as a long-term investment in payoffs that can be huge. For example, as you read in this chapter, a scientific inquiry into how babies learn their first words translated into a radical new way of designing retail spaces that adapt to customers and a tool for brand managers to gain real-time insights into the audience reaction to their TV ads. These types of innovations may have never have emerged in a timely fashion from either an applied or basic research environment, but happen regularly at the Media Lab.

CHAPTER 5

The New Normal

"Cool!"

Sam K. grins as Micah Eckhardt, a graduate student in the *Affective Computing* group, hands him a small tablet computer. On this hot, sunny August afternoon, Sam is wearing shorts, a t-shirt, and a red Astros baseball cap that just barely keeps his mop of dark hair from flopping into his face. The thirty-two-year-old attends the day program for young adults at the Cove Center in Providence, Rhode Island, which is part of the Groden Network, the state's largest provider of services to the autistic and developmentally disabled.

The device that Sam is cradling in his hands is called *iSet*, an acronym for *interactive social emotional toolkit*. iSet, which is designed to help people with autism better interpret the nonverbal signals they both receive and send out into the world, is essentially a lightweight mobile Samsung computer attached to two tiny webcams: One faces the user, and the other faces away and is focused out on the person with whom the user is interacting. As the camera captures video of the person's facial expression, a program called *FaceSense* catalogs and analyzes every raised eyebrow, puckered lip, smile, frown, and nod—then reports back to the user whether the person with whom he or she is speaking appears to be happy, sad, angry, interested, bored, confused, or something else. At the flick of a switch, iSet will reverse its focus and tell the user what his or her own facial expressions are conveying to those around him or her.

iSet was developed by research scientist Rana el Kaliouby, in collaboration with her colleague and advisor Professor Rosalind Picard, who is head of the lab's Affective Computing group. Affective Computing, a field that Picard pioneered, is devoted to designing computers and software that can detect, measure and respond to human emotions. Today, a team made up of el Kaliouby, Eckhardt, and Matthew Goodwin, the Media Lab's director of clinical research and associate director of research at the Groden Network, is busy putting iSet in the hands of the people for whom it was originally designed: teens and young adults who, like Sam, struggle with autism, Asperger's syndrome, or other disorders that make it difficult to recognize, process, or respond appropriately to social and facial cues.

When it comes to autism, there are more questions than answers. No one knows what causes it, and there is no effective treatment or cure. Even though autism appears to be on the rise, no one is sure how much of the spike is caused by greater awareness on the part of medical professionals and more aggressive and earlier diagnosis. In any case, about 1 out of 100 newborns will eventually be diagnosed with some form of autism by the age of eight, compared with 1 out of 150 just a decade ago. Currently, 670,000 American children and young adults under the age of eighteen have been diagnosed with an autism spectrum disorder. There is also a sizable adult population that is harder to document.

As you will soon see, iSet isn't just another assistive tool for people with autism. It's part of a technology revolution that will forever alter our most basic notions of human abilities and disabilities, of what it means to be "normal." And Sam isn't just another test subject for a field study. He is a "lead user" of technologies that have the potential to make *everyone's* life better, to transform many aspects of our society, and to create amazing new opportunities for health and business along the way.

WE'RE ALL DISABLED

Six months after I arrived at the Media Lab, we held our summer 2006 faculty retreat at a beach resort on Cape Cod. During a break I was taking a stroll around the grounds to get some fresh air and stretch my legs when I spotted Professors Seymour Papert and Marvin Minsky. They were chatting at a table alongside the swimming pool, and I wondered to myself what these giants of computer science, pioneers in the field of artificial intelligence, could possibly be talking about. So I deliberately wandered over, and they invited me to take a seat.

The subject of their conversation was the twenty-minute film we had seen first thing that morning entitled *Mountains without Barriers*. It documented a festival that had been held the summer before in the Italian Dolomites that brought people with physical disabilities together with those who create the technology to assist them. The film begins with what appears to be a typical mountain climbing scene, but we soon learn that we are actually watching a double amputee, the Media Lab's own Professor Hugh Herr, leading two blind climbers up a nearly vertical 2,000-foot rock spire. The ascent went without unusual incident, but when they reached the summit, Herr and his climbing partners were dramatically evacuated by helicopter due to a suddenly approaching storm, Herr leading the way.

The three of us were very impressed by this story—not just with the intrepid climbers but also with the amazing assistive technologies that were featured in the film. This was a good opportunity to ask two of the brightest people on the planet a question that had really been bothering me. I really loved the research that was happening at the Media Lab to help people with disabilities—that is what attracted me in the first place—but how could we get our corporate sponsors to be more enthusiastic about this work so we could afford to expand it? After all, most businesses had long considered the "disabilities market" to be too small to get their attention.

Papert didn't hesitate for a moment. He shot back that I was looking at it all backward. His extensive research into learning disabilities had caused him to believe that we are *all* disabled, just at different levels. For example, people with autism have difficulty reading the emotions of others, but so do most of the rest of us on occasion. Alzheimer's patients struggle with memory problems, but who among us doesn't at some level, at some time or another? Amputees have trouble walking normally, but so too do many elderly people. He couldn't see why technologies that help people considered to be "disabled," both mentally and physically, wouldn't have a huge market—virtually everyone on the planet.

This was an amazing insight! It got me thinking about how to best communicate Papert's ideas to a broad audience, including our sponsors. I thought about this during the retreat, and before it was over, I e-mailed John Hockenberry, an Emmy Award–winning journalist who had collaborated with the Media Lab years before, to get his thoughts. Hockenberry happens to have a physical disability—he is a paraplegic as a result of a spinal cord injury he suffered as a youth. He was intrigued by the concept, and I arranged a visiting appointment for him at the lab. By early winter he had partnered with Professor Hugh Herr to co-produce and co-host a major conference that would serve as a launch pad for the Media Lab's next chapter. Its theme was how the coming convergence of humans and machines will enable us to "upgrade" our minds, our bodies, and even our identities, focused on people who are traditionally called "disabled" but then extending to everyone in society. They called the conference "Human 2.0," and it was scheduled for our spring 2007 Sponsor Week.

It was a day I will never forget. Almost a thousand people crowded into the Kresge Auditorium, and our corporate sponsors were seated next to academics and other interested parties from throughout the world, some of them in wheelchairs or modified Segways, and others sporting experimental prostheses. Hockenberry, who emceed the day, kicked things off by bringing two unlikely props onto the stage. The first was a manual typewriter, the one he had used in battle zones as a war correspondent and that he jokingly described as "a laptop that prints while you type." He informed the audience that the typewriter had originally been developed to enable the blind to write. The second was an old-fashioned rotary phone, which Alexander Graham Bell had first invented to help the deaf communicate. But this was more than just a fun bit of theater; it drove home the point that technologies invented for the disabled could well open up huge new commercial market opportunities in the coming decades.

Hockenberry crisscrossed the stage deftly in his modified Segway as he introduced the day's featured speakers, which included neurologist and author Oliver Sacks, architect and designer Michael Graves, and athlete and supermodel Aimee Mullins, all with fascinating perspectives on the power of technology to enable them and others to transcend disabilities. Also on the agenda were half a dozen Media Lab professors, three of whose stories are presented in this chapter.

"AFFECTING" CHANGE

The Cove Center at Groden is a bright, airy, three-story beige stucco structure that was built on the former site of a gear factory. From the second-floor window, Sam looks wistfully toward the picnic tables in the center's outdoor courtyard, where a half-dozen of his peers are eating lunch under the shade of colorful umbrellas, next to the two large metal gears that are displayed like sculpture. Sam is clearly torn between wanting to be outdoors with his friends and staying indoors to work, but when given a choice, he decides to stay inside with the Media Lab group. After all, there's a lot to do. He and some other young men and women at the center had been helping the Media Lab team improve on an earlier iteration of iSet, which was a clumsy getup with separate cameras that had to be worn around the neck or attached to the user's clothing. Overall, he and the other students liked the device, but they recommended that the researchers figure out how to embed the cameras into the computer to cut down on the size and bulk. This wasn't easy to do. Video gobbles up enormous computing power, and so does the program that runs iSet, which places a heavy burden on the computer's battery. So it was a major accomplishment when the group designed a smaller device that could run both simultaneously.

Today, Sam is testing the new streamlined version of iSet, which runs on the latest ultra-mobile PC donated by Samsung. Micah Eckhardt, a graduate student in the *Affective Computing* group, instructs Sam to point iSet at Matthew Goodwin, who is standing a few feet away. Goodwin flashes Sam a big smile, and within seconds, a small colored bubble appears on the screen, displaying the word *SMILE*. The bubble grows larger as the system grows more confident in its assessment.

"The computer says that Matthew is smiling. Is it right?" asks Eckhardt.

Sam looks at Dr. Goodwin and then back to the screen. He nods in agreement, "Yes, Matthew is smiling."

Eckhardt flips a switch to activate the camera that is facing Sam, who has to turn his red Astros cap backward so the camera can get a full view of his face. He tells Sam to smile, and Sam obliges.

Eckhardt asks, "Is Sam smiling?"

"Yes," Sam answers, and his grin gets even bigger when he sees the *SMILE* bubble pops up on the screen.

"Cool!" Sam exclaims again.

Autistic people, as Goodwin puts it, are not "social conformists." They can be disarmingly direct because they lack social filters. Much of this has to do with the fact that they have significant trouble reading the facial cues telling them when they are making others feel uncomfortable.

Sam is the perfect example. He is charming and considerate one minute, inappropriate and awkward the next. For instance, after noticing that postdoctoral student Rana el Kaliouby is wearing a yellow and white floral hajeb, the traditional head scarf worn by Muslim women, he advises her to avoid the bacon, lettuce, and tomato sandwich in the center's cafeteria (in deference to the her religion's prohibition on eating pork). But then he launches into a lengthy monologue describing the shows he likes to watch on the Discovery Health channel, and he does so in increasingly graphic detail while his audience begins to shift uneasily. At that point, Goodwin gently intervenes, telling Sam that it is too close to lunchtime to talk about such things. One day, when the technology has advanced to the point that Sam can carry iSet around with him, it will alert him when people are put off by or losing interest in what he is saying. Or better yet, Sam will have learned these cues from iSet and will no longer need it.

Goodwin, who divides his time equally between the Groden Care Center and the Media Lab, notes that much of the research being done on facial expressions is based on anecdotal or observational data, and if quantitative is confined to artificial laboratory settings. In contrast, the iSet work the team is doing at Groden marks the first time that objective data about human facial cues has been gathered in the course of daily life. He explains, "I can analyze things that have never been examined scientifically outside of the laboratory, in real-world context, like how long do you look somebody in the eye? Too fleeting can be interpreted as rude, too long may considered combative. If we're going to develop technology to teach people this skill, we need to know the appropriate length of time for normal eye contact." This approach—what he calls *measurement in the wild*—is the same one Deb Roy took in his

Speechome project, and in both cases it relies on inventing technology that fits effortlessly and seamlessly into the fabric of a person's daily life.

The inspiration for iSet dates back to 1997, when Rana el Kaliouby was a graduate student at the American University in Cairo. As she and her then fiancé, Wael Amin, drank coffee in a nearby café, Rana casually mentioned that she was interested in doing her master's thesis on how technology could transform people's lives. Wael noted that he had just read a review of a book that touched upon similar themes, and he offered to order it for her. A few weeks later, he handed her a copy of *Affective Computing*, by Rosalind Picard. El Kaliouby was so taken with the subject she decided to develop and ultimately build a facial recognition system that could teach computers to understand emotion—like the one that Picard had described in her book.

When Picard first joined the Media Lab in 1991, she was the last person her colleagues thought would be breaking down long established boundaries. With undergraduate, master's, and doctoral degrees in electrical engineering, she suspected that in those early days some of her colleagues thought she was too staid for the freewheeling Media Lab culture. "I think they thought I was too much of an algorithm-writing, number-crunching *engineering type*," Picard recalls with amusement. Of course, that was before she questioned the validity of forty years of research in the field of artificial intelligence (AI) and created an offshoot that directly challenged AI dogma. As it turned out, her bold ideas and approaches drew from so many fields and areas of research that she had to develop an entirely new discipline around them.

Artificial intelligence is the branch of computer science devoted to making smarter machines. Since its inception in the late 1950s, much of early AI research had been focused on trying to duplicate the neural architecture of the human brain to produce a machine brain that would be exactly like the real thing, only better and faster. "We really didn't know that much about the brain at that time, but we did know a lot about human vision, how the signal goes from the thalamus to the cortex and so forth, and how all these things get processed, and our goal was to try to build a computer architecture that would duplicate that design," Picard explains. But the more Picard, who was at the time working with a team at Bell Labs that was trying to build the ultimate smart machine, studied how the human brain worked, the more she began to wonder whether their attempts to "build brains on chips" would ever pan out. She had become interested in the work being done in a new field called *affective neuroscience*, which focused on the role emotions play in how the brain processes information, and in how we make decisions. She began to question whether it was even possible to approximate human thinking and decision making in a machine without factoring in the element of emotion, which was ignored in conventional AI circles.

Luckily, the time for these kinds of questions was ripe. The field of neuroscience was flourishing, and researchers were finding emotion to be so key to rational thought that some were proposing radical new theories about its importance in all areas of our lives. Perhaps the most famous of these researchers was Daniel Goleman with his influential theory that *emotional intelligence* (EQ) was even more important for success in life than IQ. This made perfect sense to Picard, who notes, "If you think about it, every decision that we make—even the simplest one—is colored by emotion. If you see a ball coming at you, you automatically duck to avoid getting hurt, but your emotions are involved in that decision. How can machines make decisions like humans if they don't factor in human emotions?"

Emotion, however, was a word that Picard would have preferred not to link with this new field she was pioneering, for several reasons. First of all, there was a time when even she would have considered it "unscientific" to discuss emotions in the context of computing. Second, the word *emotion* was often associated with the word *irrational,* and who wanted to be known as the designer of "irrational" computers? That conjured only science fiction–worthy images of miswired robots going rogue and trying to take over the world. Finally, as one of the few women in the lab, she didn't want to dredge up the stereotypical view of women as "too emotional."

So to avoid all the negative connotations associated with the word *emotions* or *emotive,* Picard dubbed the new field, one that could transform business, entertainment, psychology, health, and learning and that could even change the way humans relate to other humans, *affective computing.*

Picard and el Kaliouby crossed paths in 2004 at the University of Cambridge, where Picard remembers being amazed by el Kaliouby's demonstration of her early version of what was then called the *Mind Reader*; it made other emotion recognition technology look primitive. While the latter could only detect obvious facial expressions like happy, sad, anger, or disgust, el Kaliouby's program could decode more subtle emotional states like interested, disinterested, unsure, concentrating, agreeing, or disagreeing. El Kaliouby invited Picard, the woman whose book had changed the course of her life, to be an examiner on her doctoral thesis, and thus a partnership was born that could change the course of many people's lives.

Picard next invited el Kaliouby to join her Affective Computing group, and with a grant they secured from the National Science Foundation el Kaliouby created the first version of *FaceSense*, the novel software on which iSet is based. It's grounded in an old manual facial expression scoring system, developed by a Swede, Carl-Herman Hjortsjö, and refined by renowned psychologist Paul Ekman and several others. The system maps facial muscle movements to numbers, but these movements, like raised eyebrows, don't always tell you what a person's emotional state is. To do this, el Kaliouby borrowed a set of videos from her mentor Simon Baron Cohen of Cambridge University that showed actors, including Daniel Radcliffe of "Harry Potter" fame, making facial expressions for different emotional states: happiness, interest, boredom, concentration, confusion, etc. She then trained sophisticated machine learning models to automatically detect the mappings. As el Kaliouby explains:

You start with 1,000 videos of people portraying a rich range of emotional states with their faces. You then hire a panel of ten or twenty people who have never seen the videos before, ask them each to watch each of the 1,000 videos and choose a label that best matches the emotional state portrayed by the person in each video. You then count the number of times the panelists agreed on a label for a video, say, concentration, and if that represents the majority, you label that emotion as an example of concentration. When you are done, you feed FaceSense these 1,000 labeled videos, and it learns what facial expressions showed up most often in all examples of concentration—coming up with a facial "signature" for concentration.

Not until el Kaliouby started demonstrating the system to the corporate sponsors of the Media Lab, however, did its full potential truly come to light. The more sponsors saw the system in action, the more they could imagine ways of using it in their businesses. For example, one leading beverage company has now used it at their taste-testing facility to measure consumers' reaction to two beverages that had already been brought to market. Both had been received positively by focus groups, but one had been a total flop in the marketplace. To find out why, el Kaliouby used face reading technology to measure people's emotional responses—their real responses—to the two beverages, and then she compared their facial cues to the grades they gave the products on a traditional survey. The expressions people made in that fraction of second following their early sips—that wrinkle of the nose in distaste, or that slight curl of a mouth in delight—were more consistent with how the product had actually done in the marketplace than what people had said about the products on the survey. It certainly wasn't lost on anyone that if the company had done an analysis of the facial expressions *before* bringing the beverage to market, it might have saved a good deal of trouble and money.

iSet isn't the only technology the Affective Computing group initially developed for people with disabilities that has turned out to have much broader uses. Take *iCalm*, a wristband with sensors that monitor stress by measuring electrodermal activity in the sweat glands of the skin. It measures how frazzled you are on the inside, even if you appear to be perfectly calm on the outside.

Picard sees this technology as vital for people who have difficulty communicating their anger or frustrations until they have reached the boiling point. Many parents and teachers of autistic children have complained that they seemed to go from calm to agitated to violent in a split second. Meanwhile, some autistic people have complained that they can't understand why others can't see how upset they are getting. But in reality, the emotional storm can be brewing for a long time, yet the person's outward signs belie his or her inward state. iCalm compliments iSet because while face reading technology is good at tracking what is known as *valence*—the component of emotion that reveals how much you like or dislike someone or something—it doesn't capture what's going on beneath the surface. A child's facial expression can create the impression that everything is fine while he or she is actually seething on the inside. So unlike a facial reader like iSet, iCalm can detect the involuntary, gut reactions that could indicate a child is upset or agitated, even if his or her face doesn't show it.

But just as with iSet, iCalm soon found its way to the general population, as well. It isn't hard to see the appeal of such a device. After all, every one of us experiences some level of stress in our everyday lives. Wouldn't it be great if we had a way of knowing exactly what was triggering it, or of detecting that it was about to creep up on us? Sensors on this comfortable wristband track a person's sympathetic nervous activity, a good indication of his or her stress levels, and report the data back to a laptop or mobile phone, which can then provide the user with a daily readout. In the past, stress levels could be measured this way only while tethered to a machine at the doctor's office, but now, anyone can track his or her stress levels while going about his or her day-to-day business.

Picard herself is often seen around the Media Lab, and elsewhere, wearing the wristband as she goes about her daily activities, and she has been known to offer it to others who seem to be in need of some stress reduction. She even did this one time through a crack in the doors of an elevator in the old Media Lab building, where I was stuck between floors for over an hour with a group of ten very tightly packed and anxious executives from a major media company that was considering becoming a sponsor of the lab. Only a few took her up on her offer, and unfortunately the company didn't join the lab (they never told me if it was the traumatic experience in the elevator).

But many of our corporate sponsors are extremely interested in using iCalm in a whole variety of ways: from monitoring stress in call centers, to measuring audience engagement in movie theaters.

In fact, the response to iSet and iCalm has been so extraordinary that el Kaliouby and Picard came to me one day with a long list of autism research organizations and corporate sponsors who were asking for the technology to create pilot projects. Realizing that they could never meet these demands with just a limited group of graduate students, we concluded that the only way they could possibly bring their technologies to everyone who needed them was to build a successful company. So they formed a startup called Affectiva, Inc., to manufacture and market commercial versions of iSet and iCalm. Affectiva's initial customers were clinical researchers, schools, and therapeutic facilities who work with people with autism and other disabilities that interfere with emotional communication, but I am told that the company is also branching out to develop exciting new commercial applications for their products as well, including market research and product testing.

LIGHTING UP THE BRAIN

If you walk out of the fourth-floor elevator in the Wiesner Building, turn right through the glass doors, and wander down the equipment cluttered corridor, you will happen upon the workspace of the *Synthetic Neurobiology* group, where Professor Ed Boyden and his students are inventing tools to rewire brain circuits and repair the "defects" that cause neurological disorders like epilepsy, Parkinson's disease, and memory loss. Although the Media Lab has embraced researchers from almost every discipline over the years, adding a bench neuroscientist like Boyden, whose work literally probes the deepest recesses of living brains using technology, was a dramatic signal that we were moving aggressively into the new era of Human 2.0.

Boyden's "workshop" looks nothing like any of the others at the Media Lab. His looks more like a traditional science lab, overflowing with centrifuges, plate readers, incubators, biosafety cabinets, bacterial incubators, spectroscopy devices, and DNA sequencers, and it is populated by researchers in lab coats and goggles preparing assays and peering into microscopes. On the other hand, Boyden's group, which is unlike any other neurosciences group in the world, is, in many ways, quintessentially Media Lab. Its unorthodox agenda is driven by researchers from a dizzying variety of disciplines: computer engineering, aerospace engineering, physics, cell biology, neurology, medicine, medical engineering, chemical engineering, electrical engineering, and even literature. And these researchers aren't just inventing new technologies for reprogramming brain circuits. They are pioneering an entirely new interdisciplinary field for treating brain disorders.

That's quite an ambitious agenda, considering that so much of what goes on in the brain is still unknown or poorly understood. Although there is a great deal of ongoing research on treating brain disorders, much of it is being done in a scattered, ad hoc fashion. "One group in England makes a magnetic neural stimulator, and one group in France uses an electrode to try to treat Parkinson's, and one group at Bell Labs tries to image the brain using an MRI, but the problem is, one group can't learn from and build upon the work of the other groups in a systematic way because there are no unifying principles for using technologies to address brain disorders. That's what we're trying to do here," Boyden notes. He sees the Media Lab as such a unifying force, perhaps just because it is so unlikely a player. By making the technology he invents in his lab "open source"—that is, available to any lab or researcher who wants to use or build on it—he's taking the first step in breaking down those silos to foster better collaboration across the field. He has already distributed parts of the technology he is developing to 350 labs around the world, and more are signing up every week.

The problem isn't just the lack of cooperation among researchers, though. It's also the complexity of the brain itself. The adult brain weighs about three pounds, but it is packed with 100 billion neurons, each with 10,000 connections between them. Just a cubic millimeter of brain, the size of a small pencil point, contains hundreds of thousands of cells and millions upon millions of connections. Figuring out exactly how these billions of parts work together is a daunting enough task, yet harder still is designing tools that can target just those brain cells that are involved in particular functions, like those that respond during stressful episodes or those that quiet the sensations of pain. But Boyden thinks of the brain as a computer: If one connection breaks down and its neural calculations go awry, it should be possible to identify the exact points that need fixing and reprogram them.

Boyden's method for repairing these broken connections is to target individual neurons with ultrathin light beams, which is a departure from the conventional pharmaceutical approach that has dominated the treatment of neurological disorders for the past century. Sadly, the conventional approach has produced virtually no cures and few effective or long-term treatments for the nearly 1 billion people who suffer everything from depression and anxiety, to Alzheimer's and Parkinson's disease, to seizure disorders. Even the best drug therapies are "sloppy" because they don't precisely target the specific portions of the brain that may be causing the problem but rather are taken up by the entire brain and often affect normal as well as abnormal parts, which increases the risk of unpleasant side effects. Moreover, many of these medications don't work as well as hoped; as much as 40 percent of the effect of an antidepressant is attributed to the placebo effect, and, despite a growing and desperate need, there are no effective treatments for Alzheimer's and no cure in sight. The problem is compounded by the fact that we don't fully understand the principles of how or why many of these drugs work. We may know the general region of the brain that they are targeting, but we still don't understand exactly which neurons are actually impacted by the drugs.

"It would be as if I were trying to repair a laptop but didn't know that it had processors or displays or hard drives. I just start throwing random chemicals at it." Boyden explains further: "I could throw this bottle of ink at it and hope that it works. But if I understand that the processor is fine but we're losing files, I can guess that the problem is with the hard drive. With a computer, we have a model of how that thing works, that allows us to understand and improve our engineering of things. Not so with the brain."

Six years ago, when Boyden was working toward his PhD in neuroscience at Stanford, he and a colleague launched a late-night spare-time project that struck gold: they discovered new and ultraprecise methods of targeting specific brain cells, something that is nearly impossible to do with conventional neuroscience techniques. They began by isolating the genes responsible for producing light-sensitive proteins in special types of algae. Next they inserted these genes into a harmless virus and infected selected brain cells in mice with the treated virus. The mouse's brain cells infected in this way automatically expressed the special genes to produce light-sensitive proteins. Then, they used tiny optical fibers to pulse light into the mouse's brain, which, depending on the kind of virus that was used, could either turn the cell on or off like a logic gate in a computer chip. This made it possible to program individual cells and neural circuits, so they could see precisely which groups of cells were involved in specific activities, such as those like memory or anxiety. It also opened up the possibility of using light to stimulate underperforming cells or to silence cells that are causing mischief, as in the case of epileptic seizures. This new approach has emerged as an important new field of scientific inquiry called *optogenetics,* and Boyden is considered to be one of its pioneers.

Boyden accepts the fact that the time frame for the Synthetic Neurobiology group's research to have actual impact on people might be far greater than any in the history of the Media Lab. But he believes strongly that his massive open network of collaborations with scientists throughout the world in spaces such as neuropathic pain, epilepsy, and Parkinson's disease—what he calls the *combinatorial model of research*—will dramatically accelerate the progress of his technologies toward cures. And since he is a man in a hurry, he has taken a concrete step to prove his point. Together with a professor at the University of Southern California (USC) he connected with through his open network, Boyden has founded a startup company called Eos Neuroscience, dedicated to commercializing the optogenetic methods for the treatment of retinitis pigmentosa, a genetic eye condition caused by the loss of photoreceptors in the retina that underlies a number of serious vision problems that affect millions of people and that often leads to blindness. Their approach is to use light-sensitive proteins to make the healthy neurons in the retina sensitive to incoming light, essentially turning the rest of the retina into a camera to restore vision.

But for Boyden this is only the beginning. Once he and his collaborators have advanced the science of optogenetics, the next step is to develop what he calls *brain co-processors,* a new project funded by Microsoft co-founder Paul Allen and by Media Lab sponsor Google. These are essentially real-time computers that can accept as input signals from millions of points in the brain, run diagnostic algorithms that can understand what is malfunctioning, and issue commands to these points to counteract the defect.

When Boyden describes this in his typically dense scientific jargon, what I envision is an iPod for the brain, attached to a fashionable brain interface baseball cap and running hundreds of "neuro-apps." Early versions will address specific brain disorders. But in the spirit of the Human 2.0, Boyden also envisions that such brain co-processors can ultimately be designed to counteract the neural deficits that every one of us encounters throughout our lives, such as addictions of various types, occasional bouts of depression and memory loss as we age.

The new Media Lab building, designed by Pritzker Prize–winning architect Fumihiko Maki   Andy Ryan Photography

Andrew Marecki, of the Biomechatronics group, performing a metabolic test of a 2011 version of the running exoskeleton   Andy Ryan Photography

Hugh Herr, head of the Biomechatronics group

Len Rubenstein

Riley Crane, postdoctoral fellow of the Human Dynamics group, explaining how the MIT Media Lab team won the DARPA Red Balloon challenge   Kris Krüg

The third floor atrium of the new Media Lab, at the center of a building that is as open and transparent as its research   Andy Ryan Photography

Amy Farber, founder and CEO of the LAM Treatment Alliance, who partnered with the Media Lab to develop a new method that allows patients with a rare disease called LAM to fully participate in the search for cures   Lara Kimmerland, with the permission of the LAM Treatment Alliance

"ZeroCar," an early test platform for the Smart Cities robotic wheel, a key component of CityCar, with student Laura Aust   MIT Media Lab, Smart Cities

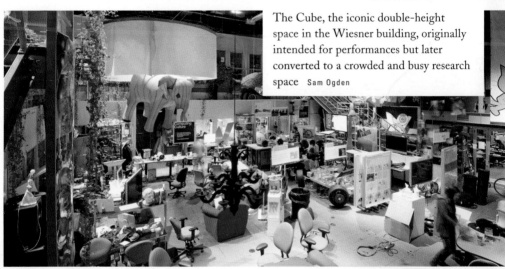

The Cube, the iconic double-height space in the Wiesner building, originally intended for performances but later converted to a crowded and busy research space   Sam Ogden

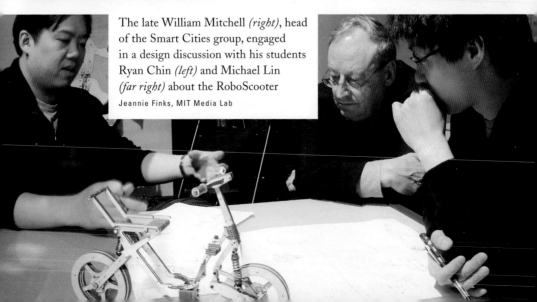

The late William Mitchell *(right)*, head of the Smart Cities group, engaged in a design discussion with his students Ryan Chin *(left)* and Michael Lin *(far right)* about the RoboScooter   Jeannie Finks, MIT Media Lab

Pranav Mistry, of the Fluid Interfaces group, inventor of SixthSense   Sam Ogden

Pattie Maes, head of the Fluid Interfaces group and cohead of the Media Arts and Sciences program

Andy Ryan Photography

A mobile phone keypad as "handy as your hand" when using Pranav Mistry's SixthSense

L. Barry Hetherington

A view of the large central workshop in the new Media Lab   Andy Ryan Photography

The first protoype of Drawdio, an electronic pencil that allows users to "draw" and play music   Jay Silver, MIT Media Lab, Lifelong Kindergarten

Tom Lutz *(left)* and John DiFranceso *(right)* in the communal workshop of the new Media Lab in a rare humorous moment   Andy Ryan Photography

A CityCar half-scale prototype, in folded position, with Smart Cities researchers *(left to right)* Lars Imsdahl, Ryan Chin, and Nicholas Pennycooke Sam Ogden

Prototype of the Robotic Wheel, developed by the Smart Cities group for the CityCar
Raul-David "Retro" Poblano, MIT Media Lab, Smart Cities

Conceptual rendering of a future CityCar parking and charging station in New York City
William Lark, Jr., MIT Media Lab, Smart Cities

Deb Roy *(center)*, head of the Cognitive Machines group, discussing activity trace from Speechome with Ronnie Kubat *(left)* and George Shaw *(right)*
Andy Ryan Photography

A single Speechome video frame taken from the ceiling-mounted camera in the kitchen of Deb Roy's home   MIT Media Lab, Cognitive Mechanics

A thirty-minute activity trace from Speechome of Deb Roy and his son as they play in their living room (Roy's trace is in green and his son's is red)
MIT Media Lab, Cognitive Mechanics

Tod Machover, head of the Opera of the Future group, with Peter Torpey and Elly Jessop, tinkering with a robotic chorus member from Machover's *Death and the Powers* opera before its debut in Monaco Sam Ogden

Joe Paradiso, head of the Responsive Environments group, modeling a wireless inertial measurement unit of the type used to monitor Red Sox players Jeannie Finks, MIT Media Lab

Performer Penn Jillette using the Media Lab's famed spirit chair to "conjure spirits" in a Las Vegas stage show Webb Chappell

Two students with autism at the Groden Center using iSET during a collaborative learning lesson to identify and understand facial expressions, with Matthew Goodwin, head of clinical research at the Media Lab, and Daniella Aube, research coordinator at Groden   Andy Ryan Photography

Left to right: Michael Chorost, Aimee Mullins, Hugh Herr, and John Hockenberry discuss the future of prosthetics at the Human 2.0 Conference   Webb Chappell

Rosalind Picard, head of the Affective Computing group, demonstrating social/emotional face-reading technology on the stage of the Human 2.0 Conference Webb Chappell

Ed Boyden, head of the Synthetic Neurobiology group, introducing his research on noninvasive brain interfaces at the Human 2.0 Conference

Webb Chappell

Cynthia Breazeal, head of the Personal Robots group, with prototypes of "the huggable" robot   Sam Ogden

Left to right: Cynthia Breazeal, Soni Chernova, Kenton Williams, and Jun Ki Lee preparing for a learning session with Nexi   Sam Ogden

Prototypes of Cory Kidd's Autom weight-loss coach robot in various stages of assembly in the Personal Robots workshop Sam Ogden

Sajid Sadi's "feedback fork," which helps people reflect on how fast they are eating and then change their behavior
Andy Ryan Photography

Ankit Mohan, of the Camera Culture group, demonstrating how to use NETRA to conduct a simple self-exam for refractive eye disease Andy Ryan Photography

John Moore, MD, of the New Media Medicine group, demoing an HIV disease-state simulator for use by patients to increase medication adherence, with David Sengeh of the Biomechatronics group   Andy Ryan Photography

Rosalind Picard, head of the Affective Computing group, demoing the Cardiocam "medical mirror" for a visitor   Andy Ryan Photography

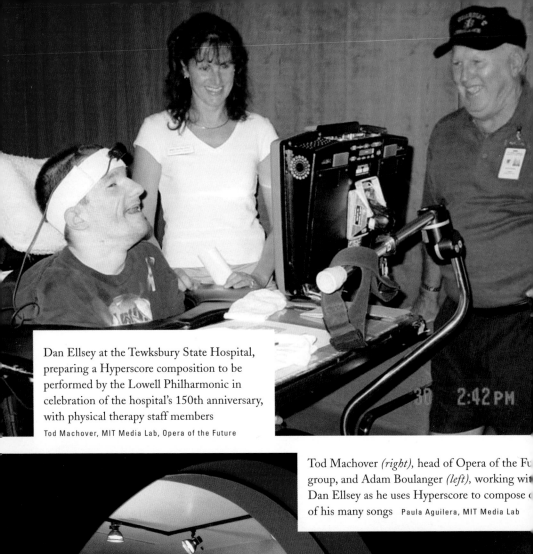

Dan Ellsey at the Tewksbury State Hospital, preparing a Hyperscore composition to be performed by the Lowell Philharmonic in celebration of the hospital's 150th anniversary, with physical therapy staff members
Tod Machover, MIT Media Lab, Opera of the Future

Tod Machover *(right)*, head of Opera of the Fu[ture] group, and Adam Boulanger *(left)*, working wit[h] Dan Ellsey as he uses Hyperscore to compose [one] of his many songs   Paula Aguilera, MIT Media Lab

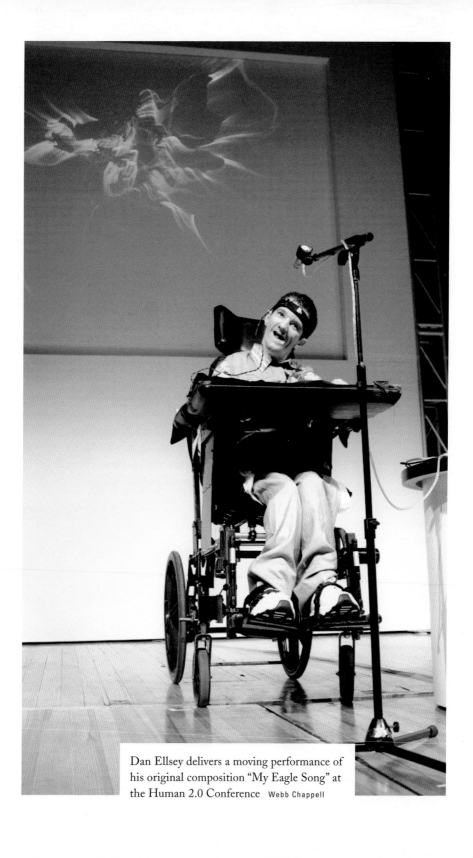

Dan Ellsey delivers a moving performance of his original composition "My Eagle Song" at the Human 2.0 Conference   Webb Chappell

Leah Buechley, head of the High-Low Tech group, demonstrating how the Living Wall can be used to turn on a lamp, play music, or send a message to a friend   Leah Buechley, MIT Media Lab, High-Low Tech

Kids using Scratch to unleash their inner programmer while Mitch Resnick, head of the Lifelong Kindergarten group, looks on enthusiastically   L. Barry Hetherington

I WALK, YOU WALK, WE ALL WALK

In December 2008, my eighty-nine-year-old mother took a fall, broke her hip, and joined the ranks of the 1.8 million Americans over age sixty-five who are injured in falls each year. For many such people, a fall marks the beginning of a nightmarish downward spiral of further injury, serious illness, and frequent hospitalization. In my mother's case, we were lucky. A timely hip replacement surgery and a six-month re-habilitation program enabled her to buck the odds. But the experience got me thinking. One afternoon, as I sat in the rehab unit cafeteria waiting to see my mother after a session, I sent an e-mail to Hugh Herr that read: "Hugh, if only we could eliminate falls in the elderly, particu-larly as the population ages, it would contribute more to improving the human condition than all of cancer research has to date. You HAVE to figure this one out."

Herr quickly shot an e-mail back to me: "Skiing right now. Will get right on it later."

Designing a mobility device that is safe and sturdy enough for the elderly is a serious challenge, one that is made even more daunting by the fact that we know very little about the biomechanics of locomo-tion. The seemingly effortless (for most of us) act of walking is actually one of nature's most elegant and intricate designs, perfected over mil-lions of years of evolution. It is an exquisitely coordinated activity, the result of hundreds of muscles working in concert to move us forward, sideways, backward, and even up and down, while still maintaining sta-bility. Furthermore, sudden movements or shifts in terrain require in-stantaneous adjustments in stance and gait, or we'll lose our balance. So any mobility device must be at least as "smart" at sensing and adjusting to the constant changes in terrain, such as when we step down from the sidewalk into the street, or up into a moving bus, all without losing our balance.

The study of human locomotion is the perfect example of how the process of developing technologies to serve the "disabled" population—in this case amputees and people with impaired biological limbs—can lead to breakthroughs that apply to everyone. As Grant Elliott of Herr's *Biomechatronics* group explains, "We're still trying to figure out how all these things work, how the spine determines when you should switch gaits. If you disturb the body in some way, we really don't know how the body is going to respond. Ironically, one of the best ways that you can learn how the body works is by observing what doesn't work." Much of what the group will learn about walking and gait will come about through the process of developing new prostheses in close collaboration with the amputees who use them, including Herr.

Herr's work developing lifelike robotic leg prostheses goes back to 2002, when he was a researcher at the aptly named MIT Leg Lab. At the time, the team was working on a project called the *Rheo Knee* in collaboration with Ossur, the Reykjavik, Iceland–based manufacturer of prostheses and orthoses. The Rheo Knee employed on-board sensors to anticipate the user's movements and special software to control the damping of the joint and adjust knee function depending on whether the person was walking, running, crouching, or doing something else. This allowed for a greater range of motion and a more natural feel than previous knee prostheses. One of the initial problems with the Rheo Knee that cropped up when it was first tested on amputees was that it actually worked *too* well; the amputees had a tendency to hesitate with each step, holding the walking leg up in the air just a fraction of a second longer than necessary, because they assumed that if they moved too quickly, the Rheo Knee, like standard prosthetics, would collapse and they would fall. So the Rheo Knee actually had to be slowed down until users came to understand that they could trust it not to give way. This led Herr to the important realization that the communication, or "human-machine interface," between the prosthesis and its wearer was as important as the prosthesis itself.

In 2004, Herr, who had by then moved across campus from the Leg Lab to the Media Lab, broke ground on his flagship project: a smart ankle-foot prosthesis called the *MIT Powered Ankle*. The young researchers knew that the test subject for their demos would be the most discriminating of users, such as Herr himself, who had a long list of requirements. Not only did the foot have to be lightweight and comfortable but it also had to be smart enough to anticipate the needs of its wearer, sturdy enough to function in all weather and on all terrain, lifelike enough to produce a natural walking motion, and efficient enough to run for at least 4,000 steps between charging. It also had to eliminate some of the things about wearing prosthetic feet that people often disliked but that most designers don't always consider to be important enough to address. For example, one of the major complaints among prosthetic wearers is that the devices snag on the bottom of their pants, tearing the fabric. This may not be a life or death issue, but it certainly is one that can be extraordinarily annoying to the wearer.

The device also had to be able to literally "think on its feet." By necessity, much of the "brain" work involved in walking and running is actually performed locally, by the feet themselves. Nerve signals simply don't travel to the brain fast enough to do the job. It takes about 50 milliseconds for a signal to travel from the brain to the foot, about one-third the time it takes to blink an eye. This may sound fast, but in fact, it's too slow to coordinate the typical running step given that the foot is in contact with the ground for less than 300 milliseconds. As the natural foot is smart by design, so must be the prosthesis, which means it has to replicate the local controllers found in the human spinal cord and limb, and it has to have the human foot's ability to produce energy. In other words, it has to have its own motor and an energy supply that can deliver the power of human muscles.

Herr was asking a lot from his students. Then he asked for more. The ankle, he decided, would not only have to be comfortable and durable but it would also have to have a self-contained battery unit that could be charged easily, anywhere, any time. Why? Because Herr liked to go climbing for days at a time and he didn't want to run out of juice.

For the next two years, under Herr's supervision, four full-time graduate students and an assortment of undergraduates each tackled a different design problem, and with every new prototype, the ankle got sleeker, smarter, and more powerful. By the time the Human 2.0 Symposium in May 2007 rolled around, the MIT Powered Ankle was ready for its public debut. In the next to last session of the day, Herr set the stage by showing a video of an amputee happily sporting a version of the MIT Powered Ankle that required him to wear a bulky backpack containing all of the computers and power supply necessary to run the foot. But in the weeks before the conference, the Biomechatronics team worked furiously around the clock to miniaturize all of this equipment and build it into the envelope of the foot. When the video concluded, Herr casually leaned down and rolled up his pant leg to reveal the MIT Powered Ankle for the first time totally self-contained in Herr's shoe and pant leg. After striding about the stage for a bit, he did the same for his other pant leg, revealing a second Powered Ankle, to thundering applause.

But although his students were making amazing progress, Herr realized even before the Human 2.0 event that their time and resources were extremely limited. Although the Media Lab was extremely adept at idea generation and building prototypes, it was not equipped to do the type of product design, manufacturing, distribution, and support required to meet the urgent need for the device in a timely fashion. This would require the resources, focus, and skill set of a commercial enterprise. At first blush there might not seem to be a huge market for foot prostheses, but in reality each year in the United States, there are 125,000 new amputees, of which 50 percent are below the knee. Between them and the steady stream of existing amputees who must purchase new devices when their old ones wear out each year, there are about 72,000 potential users that would need a new Powered Ankle each year in the United States alone.

Herr wanted each and every one of them to have access to the same technology that he did, so in 2006 he licensed the core intellectual property for the Powered Ankle foot prosthesis from MIT and founded a company that he called *iWalk*, which would operate in parallel and independently of his Media Lab Biomechatronics group.

iWalk's office, on the third floor of a Kendall Square high rise, looks a lot like Herr's Biomechatronics lab, and it is run with a very similar ethos, passion, and intensity. Most of the action takes place in the large, bright workshop where a short staircase leads to a 10-degree ramp that runs the entire length of an exterior wall. This is where Herr and his team demo and test the robotic prosthetic foot, which they are marketing under the name *PowerFoot BiOM*. Since iWalk's inception in 2006,

Herr has tested countless prototypes of the PowerFoot by walking and running up and down the stairs and ramp, as have the other amputees the group has invited in to help them refine the design. During each test, the PowerFoot unit is monitored wirelessly by a base station in the corner of the room that can check all internal functions while tracking how it adapts to changes in incline and whether it is making the necessary adjustments when the wearer goes up and down steps.

In the spring of 2008, Rick Casler, a veteran of the medical and industrial robots industry, joined iWalk as its VP of Engineering. He understood that people who use it count on it just as they do their cars. So the device would also have to work on all terrains, from snowy sidewalks to dirt roads; in all weather conditions, from 25 degrees below zero in Minnesota winters to 110 degrees in the Arizona shade; and on all surfaces, flat, hills, and stairs. And most importantly, it would have to be 100 percent reliable. "If you put on your ankle in the morning and it's balky and it won't work, you can't call a friend to drive you to the office or AAA to give you a charge. The wearer of this device doesn't have a spare, so it has to be more reliable than a car," says Casler.

The iWalk PowerFoot is not only more durable and reliable than conventional prosthetic limbs but it is also actually more powerful than the real thing. On level ground, when wearing a PowerFoot on each foot, Herr can walk faster and longer while expending less energy than a typical person does when walking. Moreoever, unlike other prostheses, the PowerFoot produces a gait that looks so normal and natural that it would be impossible to distinguish between an amputee and a nonamputee. This isn't just an aesthetic issue. It is a medical one. Gait problems can cause imbalances in the body that can result in further pain and injury. The PowerFoot BiOM was released commercially by iWalk in late 2010, and it not only gives lower leg amputees the gift of normal walking again but it also lets them more fully enjoy running, climbing, and even skiing.

The PowerFoot represents a significant breakthrough in the world of prosthetics, but only a small step toward a world without physical disability that Herr has set as his personal lifelong goal. But immense technological hurdles remain before Herr's Human 2.0 vision for a totally seamless merging of human and machine can become a reality. For example, significant advances in the interface between human nervous systems and silicon-based systems are required if Herr is to realize his dream of walking barefoot on the beach and feeling the sand between his prosthetic toes. But the Biomechatronics group already has already created prototypes that interface the electrical activities of residual musculature to its robotic prostheses, and a direct connection to peripheral nerves is in the works. When it is ready, Herr plans to be the first human test subject.

In an interesting twist on Seymour Papert's observation that we are all disabled, Herr likes to joke that he feels sorry for people with so-called normal limbs because they can't be "upgraded" like his own robotic ones. But he also incorporates these people into his vision for his work, which includes devices such as the running exoskeleton described in Chapter 1, which he projects will soon be embedded into a fashionable pair of leggings that anyone, not just amputees, can slip on to augment their "normal" limbs. This invention could have all kinds of uses: from reducing traffic congestion—think "runners' lanes" in cities and on highways—to revolutionizing the world of professional sports— imagine "exo" leagues where all players have enhanced endurance and athletic ability. And of course, let's not forget restoring mobility and helping to prevent falls among the elderly, like my mother.

FURTHER REFLECTIONS

In the past, people with disabilities like Sam, who you read about in this chapter, were at the tail end of the technology adoption cycle. They benefited from innovative new technologies only *after* those technologies had been widely adopted by the mainstream users for whom they were first designed. But for the hundreds of millions of people around the globe who suffer from these mental and physical disabilities, and their families, this is tragic.

The Human 2.0 researchers at the Media Lab are turning that order upside down. They are focusing their efforts on developing technologies for people with severe mental and physical disabilities, where the needs are the most urgent *first,* and mainstream users second. And as we've seen in this chapter, many of these technologies do indeed find their way into mainstream adoption, and in some cases it doesn't take all that long.

Seymour Papert articulated the principle behind this new way of thinking about innovation to me that day on Cape Cod when he told me that everyone has mental and physical disabilities; it's just that those who we traditionally call disabled have them to a greater degree. Researchers at the Media Lab believe that these individuals can become the new early adopters for radical technologies that will benefit all of society.

For example, Rosalind Picard, head of the Affective Computing group, notes that everyone can learn to be better at nonverbal communication, at both sending and reading those subtle signals that convey our emotions, thoughts and desires to one another. These are vitally important to all our significant relationships, yet how many times have you thought, "I didn't know I looked so angry," or "She didn't look like she cared"? Picard predicts that within twenty-five years affect-sensing technologies, such as the face reader that Sam is using, will almost completely eliminate these kinds of misunderstandings and become as pervasive as cameras and GPSs are today.

Ed Boyden envisions that within the same time frame the optogenetics technologies that his Synthetic Neurobiology group is pioneering for the targeted treatment of brain disorders, like epilepsy and Parkinson's disease, will also be used to "fix" day-to-day brain activities that are perfectly normal, but that we would desperately like to change. Take, for example, a heavy smoker who is trying to kick the habit. A brain co-processor in his hat could monitor his brain for signs of nicotine cravings, and when he's about to go over the edge and start puffing through a pack, it could launch a targeted beam of light to turn off the offending neural circuit.

In Hugh Herr's mind, the technology that his Biomechatronics group is currently developing to help amputees like him control the movements of their robotic legs by signals from their peripheral nerves and eventually their brains, could be used to control the movements of an entire humanoid robot, across the future real-time Internet. He extrapolates this to what he calls World Wide Work, a new model for global outsourcing in which human-robot symbiosis will make today's forms of telepresence seem primitive.

Herr explains that with World Wide Work technologies, humans will complement the robots' language comprehension, pattern recognition acuity, expressiveness, creativity, empathy, and even capacity for love. And at the same time, the robots will be able to complement the humans' physical strength, memory, and computational speed. The combination collaborating in tandem, either in the same room or across the globe, would perform physical work and intellectual operations much more effectively than a human or autonomous robot could perform while acting independently. It is easy to imagine that this will take things like remote surgery and dangerous security operations to a new level. But I like to imagine it could be used in all kinds of other ways as well; perhaps a craftsman in Cincinnati could work with a robotic partner in Calcutta to build custom cabinetry for a home there; or a chef in Beijing could work with a robotic partner in Boston to prepare a ten-course banquet in my kitchen.

CHAPTER 6

Living and Learning Together

Nexi tilts her head inquisitively to one side, glances down at her willowy hands, leans forward ever so slightly as she bats her baby blue eyes, and speaks in a soft, pleasant, feminine voice.

"When everyone plays with Huggable, I feel left out. When was a time that you felt alienated or left out?"

The woman she is talking to nods sympathetically, "Yes, my parents definitely like my brother best."

"Ha! That's the most common answer," says Jun Ki Lee, a researcher with the *Personal Robots* group, as he looks up from his computer screen and chuckles.

The woman with whom Nexi is chatting is Professor Cynthia Breazeal, whose Personal Robots group is in the vanguard of the movement to design a new generation of robots—like Nexi and Huggable—that can relate to people in decidedly human terms. The ultimate mission of Breazeal's group is to build robots that can live, learn, work, and play among us—and not just because such machines would be fun to have around. These robots could play a real role in improving the quality of life for the sick, the disabled, and the elderly by serving as helpful companions in the home, in schools, in hospitals, and just about anywhere they are needed.

For the last century, humanoid robots played a role in practically all imaginings of a postmodern, futuristic society. From the hilarious to the sinister, often running amok and behaving idiosyncratically, they have been a favorite subject of literary, TV, and Hollywood storytellers for generations. Robots have been portrayed as evil menaces (*Battlestar Galactica*), loyal sidekicks (R2-D2 and C3-P0 of *Star Wars*), and even unlikely saviors of all humankind (*WALL-E*). Entertaining as they may be, none of these scenarios has come to pass in the real world. And while a number of interesting prototypes of humanoid robots with capabilities that perhaps exceed even Hollywood's wildest imaginings have been built, none have made it out of the confines of the research labs and into society.

However, millions of sophisticated and costly robots *are* in use today, just not in the humanlike form we envision when we think about robots, and not always in places we commonly go in the course of our everyday lives. For example, robotic arms are becoming common fixtures in biomedical research labs, automotive assembly lines, warehouses, and operating rooms. Mobile semiautonomous robots are used to perform functions in places where it is dangerous or impossible for human beings to go. They have been used to carry out risky military and security operations, to explore the surface of Mars, and to cap the spewing oil well in the Gulf of Mexico.

Though they might not look like Rosie, the comical housekeeper from the 1960s TV animated sitcom *The Jetsons,* some robots have been put to work doing things humans would rather not be bothered to do in the home. The free-roaming, floor-vacuuming robot Roomba (made by iRobot, a company cofounded by Breazeal's mentor, distinguished MIT robot designer Professor Rodney Brooks) is by far the most widely adopted robot among consumers, with about 6 million sold worldwide as of 2011.

Still, the impact of robots on ordinary people's everyday lives has been extremely limited to date. The majority of robots in people's homes today are merely toys and gadgets, entertaining but largely impractical holiday and birthday gifts that spend most of their lifetime in the recipients' storage closets.

However, this is about to change because the barriers to widespread integration of robots in our lives and homes are about to come tumbling down. The first such barrier is cost. To function among us, even the most rudimentary robots must be able to both communicate naturally with humans and be able to sense the dynamic world around them. In the past these capabilities—including natural language processing, machine vision, and location awareness—were too costly to provide in robots designed for widespread use. But these very capabilities are the same ones now becoming standard in products such as laptops, smart phones, and digital cameras, and as a result, the economies of scale are driving down unit costs to just a few dollars. It will soon be possible to build truly capable robots that can be sold at prices on the order of notebook computers.

The second barrier is demand. Just as was the case for personal computers twenty-five years ago, robots are still awaiting the "killer applications" to ignite mass adoption. For the PC this was the spreadsheet and word processing, for the smart phone it was e-mail and text messaging, and so on. In the case of robots I believe the "killer app" (perhaps a poor choice of words) will be health and wellness. For example, in the home, robots will soon be able to play the role of diet and exercise coaches, remote physician assistants, rehabilitation trainers, and medical companions for the elderly. In hospitals, they will be used to comfort and reduce stress for patients, much the same way that dogs and cats are used in pet therapy. They will also be used to assist staff with interviewing and monitoring patients and to enable doctors to "jack in" to make personal bedside visits to anywhere from anywhere, and much more. By the end of this decade I believe there could be more than a billion robots—in homes, hospitals, and elder care settings—helping people to cope with chronic disease more effectively, live independently and with dignity as long as possible, and change their behavior in ways that will help us prevent disease and injury in the first place.

But before this can happen, one final barrier needs to be overcome. We have to design and build robots that have the social skills and common sense required to forge meaningful, long-term, and truly personal relationships with humans. Only then can they live among us and avoid that ultimate indignity they all too often endure today—being banished to the storage closet forever.

ROBOT KINDERGARTEN

Nexi, her nemesis Huggable, and the other robots that occupy the fourth-floor workshop of the Personal Robots group are essentially students in a preschool for robots. Their teachers are Breazeal's students, who are also their designers, builders, and programmers. But that's not the long-term plan. Until now, the only way robots could be taught new skills and behaviors was through the efforts of the software gurus who program them. That has been a serious limitation because the process of learning is fundamental to the acquisition of natural social interactions between robots and people, just as it is between people. That is why a key goal of Breazeal's group is to invent a new class of robots that can understand and learn a wide range of behaviors, concepts, and useful skills from average people—not just a select group of computer programmers—in the very same way that people learn from interacting with each other.

Of course, this is no small feat. The process by which people teach and learn from one another is incredibly complex, involving all kinds of subtle social cues and signals. It's also highly variable; as we all know, different people have very different learning styles. To understand the bigger picture of how people learn and to emulate that process in hardware and software, Breazeal and her apprentices need to build a whole lot of different robots, then test them in a lot of different situations, with a lot of different people. This explains why the student body of the robot preschool is so diverse. Its students come in all sizes and shapes, some with shiny metal bodies and others furry, some active and others mothballed, some intact and others gutted for parts.

When I take guests to visit Breazeal's workshop, which is probably the most popular stop on my director's tours (almost everyone asks to "see the robots"), it is not unusual to encounter a dozen or more of Breazeal's human students, postdocs, and interns, some tinkering with odd-looking robot pieces and body parts, some glued to computer screens, others talking with robots, and still others just kibitzing with each other. The area is often littered with empty pizza boxes from the night before spread among soldering irons, sewing machines, hand tools, and circuit boards. The workshop looks to me like a cross between a Hollywood studio prop room and a high-tech startup engineering lab. When it is time to move on to the next stop on the tour, I usually have trouble pulling my guests away.

Nexi is one of the class's star pupils. At four feet tall, Nexi is about the size of a typical five-year-old, but she packs a lot of punch. She is one of the first *mobile-dexterous-social* (MDS) *robots* in the world. This means that in addition to being able to express emotions and interact socially, Nexi can move around on her motorized wheels, and she has arms and four-fingered hands capable of gesturing and manipulating objects. Nexi could, in principle, motor over to a nearby table littered with pizza boxes, open a box, pick up a slice, and bring it to you. Then she could show you with her facial expression how she feels about anchovies.

Nexi was originally designed and developed under an Office of Naval Research grant, in collaboration with other universities and MIT spinoff companies, to explore how autonomous robots can be used as part of human teams, for example, to rescue a child from a collapsed building. But Nexi also provides her creators an extraordinary new platform with which to study the complex interactions between robots and humans in a wide variety of real-life, everyday situations.

Unlike many of her robotic classmates, "Nexi looks like a kindergarten student," Jun Ki Lee observes. Lee spends much of his time teaching Nexi the same kinds of social skills that kindergarten-aged children pick up from their parents and teachers. It is an interesting relationship between man and robot: Nexi is learning from Lee how to respond to and act like a human, while Lee is learning from working with Nexi how to make humans feel comfortable with robots.

Indeed, Nexi is deceptively lifelike, and she can express an astonishingly wide range of attitudes and emotions. She can look happy, sad, confused, or bored; she can blink her eyes, arch her eyebrows, and tilt her head quizzically. Her head is about three times the size of a normal head; it has to be, to accommodate the many motors that make these motions possible. Her digital computer "brain" isn't confined to her large head, however. It is wirelessly connected to a network of computers around the room that she can tap into when she needs more information. She has microphones for ears and color video cameras in her eyes, and the 3-D depth camera embedded in her forehead allows her to follow and track human facial, head, and body movements. Her "body" is exposed hardware below the "shoulders," and she moves around on wheels instead of legs. Nexi has a laser range finder that constantly scans the room and helps her determine distances between people or objects. She can't yet distinguish one human being from another—for example, she can't tell the difference between her various teachers—but the technology to make that happen is already available, and Nexi will have it soon. Nevertheless, when Nexi looks you in the eye, you feel as if she knows you, and when she expresses an emotion—like her sibling rivalry toward Huggable—you can't help but feel for her.

It is fascinating to watch Nexi's face as she morphs from one mood to the next, and despite the fact that you know that "she" is actually an "it," and that her "feelings" are determined by a sophisticated computer code, you can't help but feel as though she's almost human. And that's exactly the point. The "sociable robot" is meant to seem like a person so it can function as an ally on the job, a learning companion in the classroom, a devoted assistant for the infirm or injured, or even a reliable emergency worker to whom you are willing to entrust your life in times of peril.

The day I pop into the workshop, Nexi is helping Lee to explore a particularly hard to understand aspect of human relationships: when and why people choose to reveal personal information about themselves to others. Over the past year, Lee, with Nexi as his able and willing aide, has been conducting a number of experiments to see what type of body language, facial expression, and level of interaction best encourage someone to open up to another person. For example, in one test Nexi remained motionless throughout the entire conversation; in another, she moved her arms as she asked questions. In another test, Nexi made steady eye contact whereas in another she'd alternately make or break eye contact at certain times in the conversation. The latest iteration of Lee's study is focusing on how humans sense that another person is engaged or interested in what they are saying, which can be used to design robots with the kind of social intelligence that would make them better conversational partners.

"Whenever Matt comes around, I'm really happy," Nexi is saying. *"I love talking and chatting with him, but I'm too shy to tell him. Is there anyone that you get butterflies with when you see him enter?"*

Nexi is referring to Matt Berlin, who, along with other members of Breazeal's Personal Robots group, has spent the better part of the past decade writing the software platform that runs Nexi and their other robotic companions. Berlin joined the group soon after it formed in 2001, and he has been working with Nexi and her peers ever since.

Now Nexi's expression suddenly looks serious and quite concerned.

"Professor Breazeal is really busy and doesn't always spend as much time with me as I would like. She travels, gives talks at conferences, and has to take care of the other robots at our lab."

Breazeal laughs and admits, "Well, I have been traveling a lot these days."

The scientific and intellectual journey of Breazeal, mirrored by a steady progression of ever-more-sophisticated robots she has constructed, is not just a fascinating personal story. It is also a chronicle of the evolution of the social robotics field itself. It began when she was ten years old and was so taken with the loyal droids of the *Star Wars* movies, R2-D2 and C-P3O, that she desperately wanted a droid of her own. Then and there, she made up her mind that when she grew up, she would build one. Raised in Silicon Valley by a mathematician father and a computer scientist mother, Breazeal received her bachelor of science degree in electrical and computer engineering from the University of California, Santa Barbara, in 1989 before enrolling as a graduate student at MIT, where she became a research assistant at the Artificial Intelligence (AI) Lab's Mobile Robotics group (the Mobot Lab) then headed by Rod Brooks, already a legend in robot design. At the time Brooks was interested in the role of robots in space exploration, and he was set on convincing NASA to adopt multiple lightweight, mobile, autonomous, and relatively inexpensive robots to explore Mars (as opposed to what it was doing, which was pouring all its resources into a few very heavy and expensive robots).

Breazeal's first project was to help develop an autonomous planetary microrover, an early prototype of the kind of mobile robot Brooks was pitching to NASA. The result was *Attila* and *Hannibal*, two insectlike *mobots* that crept along on six legs built to traverse the rugged terrain of distant planets. The fact that they resembled giant prehistoric spiders was all part of a bigger plan.

The Mobot Lab was loosely using the progression by which intelligence has evolved in species over the many millennia as a model for robot development. In other words, the researchers would start low on the evolutionary scale with robots like the Attila and Hannibal, which had insectlike bodies and intelligence, before working their way up to reptiles, then to small mammals, then to big mammals, until they felt ready to create robots in the image of nature's crowning achievement—human beings. This proved to be overly ambitious. After all, it had taken nature 400 million years to evolve tiny-brained life-forms that could only slither and crawl into the highly intelligent, complex human species that think, walk, talk, and reason. But the Mobot Lab was trying to do it practically overnight.

Luckily the game plan changed around 1993, when Brooks went on a sabbatical and toured robotic labs around the world. He was particularly excited by the humanoid robotics work going on in Japan. At that time the Japanese were becoming very concerned about the growing senior population and the lack of human resources to care for them. And with good reason: By 2050, senior citizens are going to be outnumbering children fourteen and younger worldwide for the first time in human history. The social costs in countries with low birthrates like Japan are likely to be especially high. Plus, the tech-savvy Japanese scientists and business leaders, many of whom grew up as fans of Astro Boy, the robotic boy whose comic strips and TV show had been enormously popular in Japan in the 1970s, were not only open to the idea of robots living among us but they were also actively excited about it. So they came up with the perfect solution to one facet of this population problem: Why not invent a lifelike robot who could be a companion to the sick and elderly and could perform many of the same jobs that a human caregiver performs but at much lower cost?

Brooks came back to the Mobot Lab inspired, and he immediately issued his colleagues new marching orders: Forget evolution. Let's just start building humanoid robots. That is why Breazeal moved on from designing insectlike robots that roved in space and began building humanlike robots that helped people on earth. Breazeal notes the irony that although robots have already explored the depths of the oceans, braved the arctic and volcanoes, and ventured deep into space, the real "final frontier"—the place where it would be most difficult for them to function—turned out to be in human society.

Breazeal recognized right away that building robots that simply *looked*, or even acted, like human beings was not going to cut it. In order to be effective companions for people, they would need to also be able to *interact* with humans in deep and meaningful ways. At the very least, they would have to be able to understand and respond to human cues. After all, even the family pet does that! And in turn, humans would need to be able to understand and make sense of their behavior. But Breazeal also wondered how a robot could possibly be taught everything it would need to know in order to live among human beings in the real world, where situations, personalities, and habits are constantly changing and evolving. Could a robot really be taught to constantly learn about and adapt to its changing environment, and the people in it?

Inspired by developmental psychology, Breazeal looked to the human parent-child relationship and the process by which humans develop from infants into sophisticated, functioning adults, for answers.

Decades of research in early childhood development have demonstrated that the environment in which a child is raised is absolutely critical to how he or she grows and acquires adult-level abilities. Yet it takes years of patient teaching and nurturing for children to develop adult thinking and reasoning abilities.

It dawned on Breazeal that it's therefore completely unrealistic to think that these skills could just be programmed into a robot. She reasoned that just as it does for humans, it was going to take many years, and many experiences in many different situations, for the robot to develop anything resembling true human intelligence. Undeterred, she set out to make it happen. The first humanoid robot she built together with Brooks's group was named *Cog*. A far cry from the humanlike Nexi, it was a large, metallic "thing" with crude yet movable head, trunk, and arms. Although it was designed to have the physical capabilities of a human infant and could "play" with simple toys like a Slinky, it looked more like a giant motorized Erector Set than a baby. Breazeal soon realized that Cog was simply too industrial-looking to forge a real connection with people; there was no way it would tug at your heartstrings and make you want to play with and nurture it. So she decided to build a new robot from scratch.

Completed in 2000, her creation was named *Kismet*, after the Turkish word for *fate* or *fortune*, and it was designed according to what anthropologists call the *baby scheme*. These are a specific set of proportions typical of an infant that adults are hardwired to respond to in a protective, parental way, like big eyes, a high forehead, and a pouty mouth—think of a doll or a Disney character. Although Kismet was just a disembodied robotic head and neck, that head had soulful blue eyes the size of light bulbs, blonde eyebrows, fluttering eyelids, cute orange parchment ears that could perk-up with joy, or droop in despair, and two thin orange tubes for movable lips. Programmed to be as expressive as possible, Kismet could not only change facial expressions but she also actively sought out human contact by calling out in a cooing baby voice when someone entered her field of vision or by looking downcast when she was being ignored.

Of course, it took a lot of hardware to create such a social creature: twenty-one motors to animate the neck, eyes, eyebrows, ears, and lips; three cameras for sight (two for high resolution in the eyes and one between the eyes for peripheral vision); and fifteen external computers running real-time programs that controlled voice, speech processing, vision, and social behavior. The overall effect was a major leap forward in terms of a robot's capacity for human social and emotional interaction, and it established Breazeal immediately as the leader in the field of social robotics.

Kismet attracted a huge amount of media coverage, and it immediately got the attention of Kathleen Kennedy, producer of the soon-to-be-released 2001 Kubrick and Spielberg movie *A.I.*, who promptly hired Breazeal to consult on human-robot interaction, a key theme of the movie. One of the first people Breazeal met on the job met was Stan Winston, the special effects wizard who had built the movie's dozen or so *animatronic puppets*, which are electronically controlled creations that, through the magic of Hollywood, appeared on the screen as autonomous robots. One of these characters was Teddy, a furry, cuddly teddy bear robot that was the sidekick to the movie's main character, David, ironically a robot boy on a quest to become "human." Breazeal and Winston came from very different directions—she from science and he from storytelling—but they shared the same visions: to create for the first time what Winston called a "real character." In essence, they wanted to build a living robot that didn't rely on special effects but actually *existed*, in the world of people. Breazeal invited Winston to design a real robotic version of Teddy, which she would then bring to life in the real world using the magic of artificial intelligence.

Breazeal had just moved from the AI Lab across campus to the Media Lab, where she was attracted to its anti-disciplinary approach: "I needed an environment that understands not only technology but also people and design sensibilities. It was a very logical transition to move to a place where I would be able to have conversations with people around the water cooler beyond asking questions about compilers or searching a database. These techniques are still important to what I do, but I needed all the other stuff also."

The unlikely partnership between the Hollywood special effects wizard and the MIT Media Lab roboticist soon spawned *Leonardo*, a three-foot-high original character bearing a strong resemblance to those malevolent furry creatures called *mogwais* in the 1980s movie *Gremlins*. But Leonardo ("Leo") had soulful brown eyes, puppylike ears, movable arms and hands, and was named after Leonardo da Vinci because he so embodied the intersection of art, science, and invention for which the creative genius was renowned.

When Leo first arrived at the lab in 2002, he was a state-of-the-art animatronic puppet, perfect for a Hollywood movie but not yet ready to function on his own in a real-life environment. But Breazeal and her team would soon change that. Over the next year, they added a wide assortment of sensors, computers, and electronics, transforming Leo from a sophisticated puppet into a fully autonomous robot that could visually track people and objects, understand basic facial and vocal cues, and orchestrate its many motors to convey convincing emotions and attitudes. Sensors in the ears made it seem as though Leo was actually listening when a person spoke, and sensors on the hands gave Leo a rudimentary sense of touch. Some have called Leonardo the "Stradivarius of expressive robots."

Leo was the perfect subject for Breazeal, who by now headed the Personal Robots group at the Media Lab, and her students to test some of her theories on how to teach robots social skills. Once again, she turned to early childhood development for insight. Her students first spent thousands of hours programming assorted algorithms and learning mechanisms into the robot's computer "brain," and then just as much time taking on the role of quasi-parents, playing and interacting with Leo to help him develop the sorts of social skills that children learn from their caregivers through natural interaction.

"This is one of my favorite videos," Matt Berlin, PhD, says, as he plays a now iconic video of Leo showing off his social referencing skills. *Social referencing*, which is what we are doing when we look to another's emotional reaction to help us assess a novel object or situation, is an important milestone that kicks in during the first year of life, and it is one of the most fundamental ways that humans connect to each other. In the video, Berlin is waving a Big Bird doll, something Leo has never seen, in front of him. Matt is sounding happy and excited as he says, "Leo, this is Big Bird. Can you find Big Bird?" Leo nods. "Leo, don't you think that he's cool?" Leo glances at the Big Bird and then back at Matt, who's still smiling. Only once he has confirmed that Matt approves does Leo learn that Big Bird is something good, and so his ears perk up and he happily tries to reach out for the yellow doll.

Then, Matt puts Big Bird out of Leo's range of vision and begins to wave a Cookie Monster doll in front of Leo. Leo has never seen Cookie Monster either. This time, however, Berlin is using a low and foreboding voice: "Leo, this is Cookie Monster. Can you find Cookie Monster? Cookie Monster is very bad, Leo, very bad, very bad." Leo again looks at Berlin, who's frowning, then back at Cookie Monster, and he begins to look concerned. Matt clearly doesn't like this doll, so Leo learns to be wary of it. His ears droop as he recoils in fear and pushes Cookie Monster away.

"This is a demo of a relatively simple kind of learning, but a lot is going on," Berlin explains. "He's learning the names of these objects and simple associations about the appraisal of the objects, but it's embedded in a relatively rich social interaction. Leo shares attention with you, communicates through gesture, and recognizes the emotion in your face and voice. You take it all for granted because it unfolds so naturally, in a sort of tightly coupled dance, each partner's responding and adapting to the other."

Experiments like the one Berlin described are absolutely critical for the future of human-robot interaction. Much of human communication is nonverbal, transmitted by facial expression, gesture, or body movement, so ultimately, in order to communicate with humans on the same level as humans do with each other, robots must understand our nonverbal cues and learn how to size up a situation and react to it appropriately.

While Leo was capable of quite sophisticated social interaction with humans, and Breazeal's group was learning a lot working with him, he was literally tethered to the lab and couldn't be used in real-life settings. That is why Breazeal decided in 2004 that it was time to create a new generation of sociable robots that could live anywhere: homes, schools, hospitals, elder care facilities, and so on. Thus, Huggable, Nexi, and Autom (an at-home weight-loss coach you'll meet in Chapter 7), each with his or her own look, function, and destination, were born.

Huggable is probably the most adorable robot you will ever encounter. It is essentially a smaller and more portable version of Leo, but it resembles a cuddly teddy bear. Unlike Nexi, Huggable can't navigate a fire maze and pull a child out of a burning building, but it can be held in one's arms or sit in one's lap and channel the words, feelings, and emotions of an actual human. Huggable is a fully autonomous robot, but it was really designed to function as what Breazeal has dubbed a *physical avatar*. Huggable has video cameras for its eyes and a speaker in its mouth, and, thanks to sensors in the skin, it can differentiate between when a child is tickling, poking, or hugging it.

Huggable has a number of potential uses. For example, it could provide warmth and comfort to a child in Boston waiting for a medical procedure and let the child's grandmother in Copenhagen "be there" in the form of a cuddly bear rather than a cold computer screen. But that's not all. At the same time, Huggable could also have embedded devices to monitor the child's vital signs, and it could serve as an extra set of eyes for the nursing staff. At home, Huggable could also be used to teach a child a second language, via a tutor or parent who is a mile or a continent away. Huggable is still in the prototype stage, but soon it will be piloted in the arms and laps of children across the globe.

Nexi, on the other hand, has already ventured out into the real world. Breazeal and her students have taken her to visit a number of senior centers in the greater Boston area to see how well the aging population would accept a social robot like her. Overall, the seniors have really embraced Nexi. A number of them immediately shook Nexi's hand upon meeting her, some hugged her when they became more familiar, and at least one of them even planted a kiss on her forehead. Through these visits the researchers also discovered that the more expressive Nexi behaved, the longer seniors would talk to her and the more personal experiences they would reveal. They even seemed to get rather attached; after a few visits, when one woman was asked if she could envision robots like Nexi helping seniors in their homes, she replied: "To me it's almost like something I never would have anticipated, but now I would take it very much for granted."

Breazeal and her students were surprised by one observation. It turned out that all the seniors wanted to do most was "really talk" with Nexi, just as if she were another person. But although Nexi was programmed with a fixed set of talking points, inventing a robot who can do this is no small thing. One of the biggest challenges in the field of artificial intelligence is to enable machines to "really talk," that is, to carry on the type of simple and everyday natural conversations that we take for granted. To be an effective conversational partner, Nexi will have to accumulate a great deal more real-world experience and understanding than she could possibly get from interactions with a handful of busy Media Lab students. But how?

Postdoctoral fellow Sonia Chernova has come up with a clever solution to this problem, and it involves a two-player computer game called *Mars Escape!* One player assumes the role of an astronaut, and the other the role of Nexi, and together they work to complete a mission on the surface of the Red Planet before the oxygen runs out. The idea is for the program to amass many thousands of examples of human collaborative behavior—it's effectively crowd-sourcing a rich set of interactions to build memories. The gathered memories will then be programmed, or "installed," into the real-world Nexi, where they could be refined and new ones could form from experiences in the real world.

It remains to be seen if *Mars Escape!* will provide Nexi all the experience and understanding she needs to "really talk" with humans on the real planet Earth. But I believe that Breazeal and her students are coming incredibly close to developing social robots that can provide companionship as a conversational partner for an elderly person living alone, comfort a child in the hospital, or even be a member of a human rescue team.

How close are we, you might be wondering, to having a sophisticated robot housekeeper like the Jetsons' Rosie (only less clumsy), who helps you cook the family dinner, does the dishes, and then helps the kids with their homework? Berlin concedes, "We're not there yet." But he quickly adds with a grin, "Don't be surprised to look out your window and see that your neighbor's grass is being mowed by a robot that knows not to cut toys that are left on the lawn. That's definitely going to happen soon."

IT'S JUST COMMON SENSE

Henry Lieberman, PhD, who spent a dozen years at the MIT Artificial Intelligence Lab under the tutelage of Marvin Minsky before joining the Media Lab in 1987, would agree with Berlin on both counts. He acknowledges that when it comes to building a smart robot with the same capabilities as a living, breathing human, there is still a long way to go. He vividly recalls how, when the field was just starting to get off the ground, the AI community made all kinds of big predictions about the future, with lofty visions of robots that could clean your house and mow your lawn.

But there's a big difference between the Roomba that he uses to vacuum the floors in his house and a robot that actually has common sense. That difference is what Lieberman has dedicated the last decade of his career to studying: a branch of AI called, fittingly, *common-sense computing*. In his mind it's crazy that any robot, or for that matter, any computer, cannot understand that when you cut grass, you should avoid mowing over toys left on the lawn. After all, that's the kind of common-sense knowledge that a child picks up naturally in the very early years of his or her life. But for all the talk of "smart cards" and "smart phones" and "smart homes," most computers in our lives today lack this and the millions of other pieces of common-sense knowledge about how the world works that come naturally to any six-year-old. Computers may have become indispensable to contemporary life, but they are still completely ignorant about the people they serve. Lieberman believes that if computers of all types, not just robots, could exercise the common sense of just a child, they would not only be much easier to use but they would also be much more helpful to the people who use them.

The ultimate goals of his research are quite similar to Breazeal's, but he's not just interested in building humanlike robots. He's concerned with "humanizing" *all* the various types of computers people use in their daily lives, from the tiniest embedded processors in consumer electronics products to global computing networks. Whereas Breazeal's methods of teaching robots are modeled after the way children learn, Lieberman's are different. "When it comes to teaching computers common sense, we could try to teach all of it from scratch, which is the way babies learn, but that's very difficult. Instead, we said, 'There's already a lot of common sense on the Internet among the people who are using the Internet, so why don't we simply collect it?'"

The "we" to whom Lieberman refers are a number of his colleagues at the Media Lab who have contributed to the common-sense effort, one of the longest-lived projects in Media Lab history. It is rooted in the work of Marvin Minsky, who set out years ago to create machines that were sufficiently self-aware to be able to think and solve problems with the resourcefulness of human beings.

About a decade ago, one of Minsky's protégés, Push Singh, and his colleagues came up with a clever way to enlist the public in creating a large collection of common-sense knowledge. They set up a website called the *Open Mind Common Sense (OMCS) database,* and they invited Internet users from around the world to log on to the website (openmind.media.mit.edu) and type in a simple common-sense statement, like "Elephants are heavy" or "The sky is blue" or "People go to restaurants to eat." Since the Open Mind launch in 1999, people have contributed over one million facts to the database in English, as well as hundreds of thousands in a dozen other languages. Taken separately, each statement isn't particularly revelatory, but in the aggregate, these shards of collective wisdom create a reservoir of everyday observations and experiences from which a computer can make connections and draw conclusions about the world.

Just having millions of common-sense facts in a database does not in itself make an intelligent machine. That can come about only once the computer learns how to actually *reason* with that knowledge to solve real problems of everyday life. For example, for a computer to simply know that people go to restaurants to eat doesn't automatically solve the problem "what to do if I'm hungry" unless the computer can also understand that eating makes one less hungry and thus if one is hungry one should go to a restaurant and eat. This is a part of a process called *inference,* and it begins with figuring out what information relates to the problem at hand. While human beings do it effortlessly every day, using cognitive models like analogies, it's a skill that still evades even the smartest of computers. Recently, though, Lieberman and his student Rob Speer and post-doc Catherine Havasi have invented a powerful new automated inference technique called AnalogySpace that literally reads between the lines like humans do.

From the beginning, the common-sense project was an ambitious endeavor, and many questioned whether it was possible to collect even a fraction of the accumulated common sense in the world. Lieberman estimates that today OMCS contains about 1 percent of an average person's common-sense knowledge. But he contends that given how pathetically little the various computing devices we use today know about people and everyday life, even this modest amount could go a long way. To prove this, he has focused his efforts recently on using common sense to improve the interface for smart phones, laptops, and other devices and make them much more considerate of us and responsive to our needs, and thereby much less intrusive in our lives. For example, Lieberman says that your smart phone should have the common sense to know not to disturb you when you are at the movies or in a meeting, unless of course it's an urgent call from a child.

Lieberman is also now working on the next big step for common-sense computing. He calls it *goal-oriented interfaces,* and he contends that it will soon make the millions of apps on your smart phone seem primitive. The ones we use today are designed mostly to perform individual tasks, like making a reservation at a restaurant or telling you how many calories you consumed today. True, they are becoming smarter all the time. Many already know your taste in music, movies, and books, and your location and your friends' locations, and they can predict at what establishments you might like to eat, drink, and shop. But this is not nearly smart enough for Lieberman. Rather than fiddling with these applications, he believes you should simply be able to say what you want accomplished and leave it up to the computer to figure out how to do it.

For example, let's say you are at work and you say to your smart phone, "Book me a table. I have a big date with Kim tonight." You should be able to treat it as you would a savvy personal assistant who knows you well and uses common-sense knowledge about specific choices that would best contribute to your goal of having an enjoyable date. A smart phone that also has common-sense knowledge would suggest a romantic place rather than a business venue, book it at a date-appropriate time like 8 p.m., and remind you to leave work by 6 p.m. sharp since there is heavy traffic on the Massachusetts Turnpike and you'll want to have time to change out of business attire. It would even ask you if you would also like to reserve seats for a late movie. It would show you a few alternative plans, and when you select your favorite, the phone would make the reservation, e-mail Kim with the time and location, and even ascertain whether there are low-sodium options on the menu since you are watching your blood pressure. And when that's done, it would put itself into silence mode for the evening at 7:30 p.m., except, of course, for any urgent calls from your children.

The work of Lieberman and his group has attracted considerable attention recently from a number of our corporate sponsors, including Microsoft and Procter and Gamble, and Lieberman anticipates that the first commercial products with common sense will soon hit the market. But in the meantime, I'm very excited about a prototype called *I'm Listening* that Lieberman has built in collaboration with John Moore, a PhD student in my *New Media Medicine* group. This application uses common-sense computing for a very different purpose: to enable machines to conduct medical interviews with patients in their homes *before* the visit to the doctor's office. It is not intended to replace doctors or other flesh-and-blood medical professionals but rather to identify the urgency of the symptoms and triage care and to prepare the patient and order tests in advance—all of which saves money in administrative costs, frees up time during the actual doctor's visit to talk about the things that are important to cover in person, and in some cases to even avoid unnecessary office visits, which one study puts at 25 percent of all visits.*

Imagine an on-screen avatar named Linda, a friendly physician's assistant, who through the magic of natural language processing can chat with you while you sit at home in your family room. Linda, who you'll read more about in the next chapter, isn't an actual person sitting miles away in your doctor's office. She exists only in a computer. Nonetheless, she asks you to describe the pain that prompted you to set up this virtual visit, and you say, "It felt like an elephant sitting on my chest."

---

* Karen Dunnell and Ann Cartwright, *Medicine Takers, Prescribers, and Hoarders,* Institute for Social Studies in Medicine Care Reports, Routledge and Kegan Paul, London, 1972.

She has the common sense to know exactly what kind of pain you're referring to and what questions to ask next. This response may have sounded funny, but expressing discomfort or pain using this kind of metaphor is actually the most common way that people describe their symptoms to doctors or nurses. But it means nothing to computers, that is, until now.

Using the OMCS database and common-sense reasoning tools, the I'm Listening software makes sense of the statement through the following process. First it asks itself, "What do I know about elephants?" It might first get, "Elephants are gray," but that doesn't help. Next it gets, "Elephants live in Africa and India," but that doesn't help either. But when it gets "Elephants are heavy," it can then classify the pain using the official McGill pain scale, prompting Linda to ask a followup question like, "What makes it better or worse?" If the person answers, "Walking, running, eating, or sleeping," the system will understand enough about these words to be able to figure out the cause of the person's pain.

This technology would never have existed if it hadn't been for Push Singh's brilliant research. Singh, generally recognized as one of the AI field's rising stars, was slated to join the Media Lab faculty as an assistant professor when tragically he passed away in early 2006, less than a month after I came to the Media Lab. I'll never forget the introductory chat with Singh in my office just a few days before he died. He told me that his dream was to eventually build a machine that thinks like a human, and I believe that if anyone could do it, he could have. Today, Lieberman and his *Software Agents* group are carrying on the common-sense project, and Singh's life and work continue to be their inspiration.

FURTHER REFLECTIONS

The fifty-five-year-old field of artificial intelligence (AI), which has been in and out of vogue several times over the years, is enjoying a major resurgence today. Many of the research areas presented in the past three chapters are branches of AI, and advances in other branches such as natural language processing and machine learning are making Internet searches faster and more precise, smart phones more aware, and social networks a rich source of insight into people's tastes and behavior. AI was thrust headlong into the public consciousness recently when IBM's "Watson" bested human champions in the TV game *Jeopardy!*. Watson represented a significant step forward in several domains of AI technology and, as IBM was quick to point out, has important applications in fields such as medicine, where it will help doctors better diagnose illnesses and treat patients.

Despite these recent developments, whether or not it will someday be possible to create machines that equal or even exceed human intelligence, and even what that really means, remains the subject of intense discussion and debate. The strongest proponent today is the inventor and futurist Ray Kurzweil, whose "singularity theory" says that by 2045, exponentially accelerating advances in information technology, nanotechnology, and biotechnology will result in a symbiosis of humans and machines that will allow the combination to achieve superhuman intelligence.

I thought it would be interesting to see what some of the key players in AI-related research at the Media Lab think about this question these days, so I began by asking Rosalind Picard, who, as you'll recall, invented the field of Affective Computing. There are some aspects of human intelligence that can be built, she concedes, but there are certain things that machines simply can't be programmed to do. "You can program a computer to look at the curve of a mouth, crunch some numbers, and decide that the subject is interested, an extension of logical pattern recognition, which machines can do." But, Picard adds that there is a less understood, more mysterious process that occurs when humans meet other humans—a certain *qualia* of feeling—that machines are unable to capture.

Cynthia Breazeal, inventor of sociable robots that partner with humans, maintains that the goal should definitely *not* be to make machines that are equivalent to humans, but rather complementary to them. She believes that it's actually the *differences* between machines and people that create real value. "Just think of companion animals, like a Seeing Eye dog. The dog is not human, but it brings real value to its person. It helps its owner navigate around, helps keeps her safe, and enables her to do things that would be much more challenging otherwise."

Henry Lieberman, developer of computers with common sense and disciple of AI pioneer Marvin Minsky, believes that someone will eventually build machines that think like humans, although he puts the timing at a hundred years from now. But he would prefer that we "get over" the goal of making machines that equal or surpass human intelligence and get on with the business of making existing products and service much more intelligent and much better at truly serving people.

What do I make of this? First, brain and cognitive scientists tell me that they are just beginning to understand the nature and limits of human intelligence (though we know it is a lot more than that involved when playing *Jeopardy!*), so it is really impossible to speculate if and when a machine can equal it. But like my Media Lab colleagues expressed above, I feel that rather than focusing too much energy on building machines that are as smart as or smarter than humans, we should devote most of our efforts to building machines that help us to be the best human beings we can be for ourselves and others.

I think there is no better example of this than a project that Lieberman presented, together with research partner Formspring (a popular social website for questions and answers), at the White House Conference on Bullying

Prevention less than a month after the IBM Watson extravaganza. Sadly, bullying of all types in social media has become all too common, and hard to control. Current approaches to detecting bullying, such as those used for spam detection, are ineffective since the same words that are innocent in one context might be bullying in another. Together with his master's students Birango Jones and Karthik Dinakar, Lieberman used common-sense computing to detect subtle comments in social media conversations that, in the context of the discussion, are meant to hurt. For example, a human can easily understand that "you probably ate six hamburgers for dinner last night" is meant as an insult about someone's personal appearance, because common sense tells us that "hamburgers are fattening." An application that could detect these kinds of subtleties might not be sufficient to beat champions at *Jeopardy!*, but could take a lot of kids out of jeopardy from cyber bullying that they are experiencing today.

CHAPTER 7

The Age of Agency

If you want to get a glimpse at truly radical health care reform, simply make an "office visit" with John Moore, MD, a PhD student in my *New Media Medicine* group. To find him, you'll have to navigate the obstacle course of robotic body parts strewn about the *Biomechatronics* lab, then wind your way through the equally cluttered workspaces of the *Affective Computing* group. Once you do, tucked away at the back of the second-floor research cube you will come across one of the more unusual research spaces in the Media Lab. That's because it doesn't look like a research space at all.

Here you will find two twelve-by-twelve-foot rooms, open at the front for all to see, separated by a six-foot-high partition, where we are reinventing the future of health care. The room on the left is a proto- type of a *medical collaboration space,* essentially the doctor's office of the future. Just about the only familiar fixture you'll find there is the requisite black leather examination table, which is usually located in the middle of a traditional examination room but here is pushed into a separate small adjoining room. Instead, in the center of the room is a wall of touch screens in front of a round meeting table circled by four comfortable and colorful chairs. This looks more like a teleconference space than a doctor's office.

The other room looks like a typical family room, or den, with a few couches, a flat screen TV, and a PC atop a desk. There are even a few homey touches like a potted palm tree peeking out from behind one of the couch armrests. But as you've probably guessed, this isn't just a model of typical living space. If you look a bit closer, you'll see a digital floor scale tucked away in a corner, a glucometer resting on the coffee table, and a quirky-looking device that resembles—but isn't—a clock radio sitting on the desk. What you won't see is that all of these devices, and many others out of sight, are wirelessly connected to the PC, which is linked to the screens in the room across the partition by a software platform, conceived and built by Moore, called *CollaboRhythm*.

It's a typical gray Boston day in the fall of 2010 when I come across Moore, who is taking a break from programming CollaboRhythm, relaxing on one of the couches in this New Media Medicine "family room." I sit on the other couch and ask him about his Thanksgiving holiday, but as always, he is restless and can't help but jump immediately into demo mode to show me the software's latest features. He picks up a home theater remote control off the coffee table and, after tapping a few buttons, brings to life the flat screen TV at one end of the room.

"Hello, my name is Linda, and I would like to ask you a few questions to prepare you for your visit with Dr. Moore."

Linda, the woman on the screen, who is wearing a white medical coat and horn-rimmed glasses, is not an actual person but an avatar. Technically, she is an *anthropomorphic conversational agent* who functions as a personal health liaison between you and your doctor, sort of like your own in-home physician's assistant. She looks professional, and she is friendly and warm as she asks you questions and listens patiently as you answer and ask yours. She also accomplishes much of what is normally done in the doctor's office, like taking a basic medical history, updating your medical records, and scheduling and preparing for your followup visit—everything but the exam itself. But she can do it from your home, your office, or anywhere.

Believe it or not, conversational agents like Linda have been shown to be more effective than physicians or nurses at tasks like these, most of which are conversational rather than clinical. Ironically, they may actually seem more *human* than the people working in your doctor's office, who are too busy trying to simultaneously juggle five phone calls, answer the questions of ten other patients in the waiting room, and fill out a pile of paperwork to actually pay attention to you. In the hectic environment of a doctor's office, you're likely to feel like just a nameless, faceless, neglected customer. Linda, on the other hand, is focused, soothing, and attentive. She's there to listen to you, and you alone. Moreover, she's available twenty-four hours a day, so if you wake up in the middle of the night unable to remember something your doctor told you, or you are confused about your medication, she can give you an immediate answer. Finally, you can confide in her, and in fact, you're likely to; there is overwhelming evidence that many people find it easier to reveal personal and sometimes embarrassing yet important medical information, like alcohol or drug use, sexual activity, emotional instability, or even suicide attempts, to a computer than face-to-face to their doctor.

Linda is one of a half-dozen special-purpose user interfaces to

CollaboRhythm that Moore has designed for patients and medical professionals that incorporate the latest advances in human-computer interaction technologies developed at the Media Lab and elsewhere, such as multiuse multitouch displays, gesture recognition, persuasive interfaces, real-time natural language processing, lifelike avatars, social robots, and more. If you have trouble believing that an avatar could possibly give you as thorough and probing a medical history interview as a real live person, think again. Linda not only asks basic background questions, like "What is the purpose of your visit?" and "Have there been any changes in your health since your last visit with your doctor?" but thanks to the *I'm Listening* software described in the last chapter, she also has the common sense required to ask the appropriate followup questions and offer you information you may not even have known to ask for. For example, when Moore tells Linda, "I am having bad knee pain, I guess it is my arthritis acting up," she asks, "Is it a sharp pain, or a dull ache?" He tells her that it's more of a throb. She then asks him if he is taking any medication to relieve the pain, and he responds that he is taking two ibuprofen every six hours, but he wonders if there are any over-the-counter supplements to treat arthritis. Linda promises to send him a recent article on a supplement that his doctor believes may be helpful. She records all of his questions and answers, and she assures him that she will pass this information on to his doctor.

As you may recall, the I'm Listening software also enables Linda to actually *understand* the different ways in which people typically describe their ailments and symptoms beyond just straightforward statements like "My knee hurts." For example, if you are suffering a severe kidney stone, you probably aren't going to know to say, "It feels like a kidney stone." Instead, you might describe your symptoms as "I feel like someone is sticking a knife in my back and pushing it to my groin." Linda is smart enough to infer that you are not the victim of a knife attack, but experiencing sharp, intense pain that is localized to the back but radiates anteriorly to the groin. She is able to calculate the probability of different diagnoses based on this description, your answers to follow-up questions, your medical history, and her comprehensive medical knowledge. She would then immediately contact your doctor with a summary of all of this data—listing nephrolithiasis (kidney stones) as the most likely diagnosis so that your doctor can evaluate you urgently or advise that you go to the ER. If you go to the ER, all of this data will be transmitted to the attending physician and your personal doctor will be able to collaborate virtually in the definitive diagnosis and treatment.

THE MEDIUM IS THE MEDICINE

It is no secret that our health care system needs fixing. In spite of amazing advances in medicine and therapies, costs are still astronomical, outcomes are still poor, and everyone involved is still frustrated. And while we in the *New Media Medicine* group don't claim to know all the answers, we do know one thing: The most underutilized resource in today's beleaguered health care system isn't doctors or hospitals or medical equipment. It's the *patients themselves.* We believe that by giving patients access to the right information and the knowledge of how to interpret and use it, they can become equal and full collaborators with their doctors in managing their own care—not just in the exam room during a checkup, or in the hospital bed after a procedure, but anywhere, twenty-four hours a day.

If you look closely at the effects advances in technology are having on people's ability to participate in their own care, you will see that a paradox has begun to emerge. On the one hand, for over a decade we have all been making frequent use of the Internet to search for information and connect with others about every aspect of our health and health care. Almost daily we Google some symptom, read the latest news story on what foods or vitamins or medical tests are deemed good or bad for us, or look on a drug interaction website to see if it is okay to take a certain medication. Each time we visit the doctor, we come armed with a pile of printouts from health websites, chat rooms, and blogs.

On the other hand, in the office the doctor's attention seems to be focused more on inputting, viewing, and manipulating data on the computer than on explaining what all the information means, or even listening to what we have to say. We often emerge from the visit more perplexed than before. Back home we Google the diagnosis, which is probably a futile effort given that a search for *high blood pressure,* for example, returns more than 5 million hits, most of which are at best confusing, and at worst completely erroneous. If we are dealing with a more serious health problem, there is the task of coordinating the efforts of various specialists, who seem to have incomplete or different information from one another. And the more hours we spend on websites or social media commiserating and sharing experiences with others dealing with similar challenges, the more conflicting advice we get, and the more our confusion grows.

The result is that we feel *less in control of our health, and that of our families, than ever before.* It is said that information is power, but ironi-cally all this information has imbued us with a sense of powerlessness. And we are just at the beginning of the information overload. It's now possible to get genetically tested for everything from the breast can-cer gene to the likelihood of having a child with Down syndrome, and soon, personal whole genome sequencing will be available for less than a thousand dollars. What's more, a deluge of real-time information is about to flow from the interconnected *electronic nervous system* that lives on us and surrounds us. It will soon be possible to monitor and record nearly everything about our lives (some call this the *quantified self*), including our vital signs, what we eat, how many calories we burn, what we buy, whom we speak to on a daily basis, and even the tone of our voice while doing so.

I believe that the future of health care rests on our ability to find a way to make all of this information transparent, understandable, and actionable to both doctors *and* patients, working together as a team. Only once we give people the tools to take the reins of their own care can we can finally begin to utilize our most undervalued resource—people themselves—and transform the entire system of health care in the process.

A PHYSICIAN ON A MISSION

The inspiration for CollaboRhythm goes back to 2005, when Moore began practicing medicine as a resident in ophthalmology at Johns Hopkins Hospital. He explains it this way: "Every piece of technology was disengaging me from the patient. I had to spend most of my time in front of a computer inputting information. Most of it had absolutely nothing to do with the patient. It was all about documentation for payment or for protection from lawsuits. It became pretty clear to me that we could do much more with this technology to improve the experience for both the patient and the doctor. It wasn't that we needed new technology. *We needed a new way of thinking about health care.*"

Moore enjoyed a lot of things about being a doctor, particularly the impact that he had on patients. But he quickly became frustrated by flaws in the medical system that in his mind made it nearly impossible to deliver the quality of care he felt patients deserved. Most troublesome to him was how woefully little his patients knew and understood about their own conditions and treatments. "Patients are under stress. They're often flustered, and the doctor is trying to do lots of things at once and is often rushed. Communication is poor, and patients remember only a fraction of what they are told in the office," he explains. He was so burdened by his heavy caseload that he simply didn't have the time to sit down and educate them. "You can tell a patient with glaucoma, a chronic condition that can cause blindness, that they're going to slowly go blind if they don't use their eyedrops, but they feel fine, and they see fine, so they think, 'Why fuss with the drops?'" he recalls. He could tell he wasn't getting through to them.

He didn't blame his patients. Rather, he blamed the system, which didn't do nearly enough to truly put them in charge of their own care. It was clear that simply telling patients what foods not to eat, or what pills to take and when, wasn't cutting it. They needed the tools to *understand* why they were given the instructions they were given—what exactly the drugs were doing to their bodies, what side effects they could expect, how to spot the signs that a treatment was or wasn't working, and much more. Out of frustration, he quit the practice of medicine and enrolled at the Media Lab, where today he is focused on giving them exactly that.

On that gray fall day in 2010 when I dropped in on Moore for a demo of his CollaboRhythm prototype, we leave the "family room" and walk a few feet around the partition and into his doctor's office of the future. We sit down together at the collaboration table, and I pretend to be the patient with the sore knee, while Moore now assumes the role of the doctor. He turns to the wall of screens, and with a touch transforms it from a piece of art to a collaborative view of my health. This is deliberately very different from today's examination room setup, in which the doctor is typically planted in front of his or her computer, typing his observations furiously, while the patient sits blankly and helplessly on the examination table or in a chair across the room. Instead, Moore's setup puts the doctor and patient side by side, looking at the computer screen together, like two players on a piano bench, playing a duet. Hence the name *CollaboRhythm,* for a new approach that allows doctors and patients to "play together," making the processes of information gathering, examination, diagnosis, decision-making, and education truly collaborative and harmonious.

Moore touches the left-hand screen a few times, where Linda is summarizing the I'm Listening interview that was conducted with the patient, whom I'm now role-playing, the day before in his family room. Moore notes that during the traditional, in-person previsit interview, half the time doctors and patients don't agree about the reason for the visit—a clear sign that they aren't communicating well. He also points out that patients are often distracted and tongue-tied when they arrive at the doctor's office, and as a result they aren't able to express their problems or symptoms to their doctor. Linda eliminates this issue completely, getting the doctor up to speed so that he or she is on the same page with the patient quickly, before the patient even enters the collaboration room. Moore then asks me a few quick follow-up questions both about "my knee pain" and how I have been feeling otherwise. Then he takes me to the adjoining room and proceeds to conduct a mock exam. Again I can see how different this is from a traditional office visit: Although he's still the one conducting the examination, I'm not relegated to the role of passive observer. I can follow exactly what he's doing on another screen on the wall, and I can see the results of every part of the exam as he performs it. When he takes blood pressure measurements or listens to my chest with a stethoscope, for example, I can see the readings—the same data that he is looking at—and interact with the information just as he can, by touching the screen or even by speaking.

At the same time, the system is updating my *collaborative health record* (CHR), a hybrid between a doctor-controlled electronic medical record and a personal health record, but richer in format and function than both. What's best about the CHR, however, is that *I* own all of the information in it, no matter what its source, and I can grant or deny access to any part of it, to any individual I choose. This is fundamental to Moore's premise that all data should be completely transparent to patients and that even the slightest compromise in transparency will undermine the patients' confidence as well as doctor/patient teamwork. Moore illustrates this to me in a dramatic fashion during the mock exam by showing that he literally is unable to hide any information from me, even his own notes.

We next turn our attention to another of CollaboRhythm's applications designed to involve the patient in the process of choosing medications. With this system, which drugs you take and the dosing schedules are no longer the doctor's unilateral decision. Instead, the doctor will show you the various drug options and demonstrate, by using an array of intuitive visualizations and graphs on the touch screen, what each drug actually does. This includes an animation of the biological mechanism, down to the molecular level, showing how it actually works in your body to attack the source of the disease. The doctor can then bring up a graph plotting the relationship between drug concentration in the bloodstream over time and its level of effectiveness, undesirable interactions with other drugs, and possible side effects. You even see whether the drug is covered by insurance, and with what copay, and any other information that might factor into a decision about what to prescribe. With all this laid out on the screen, you then work together, dragging and dropping information from one place to another, to select the best medication. Finally, you design a dosing schedule—that is, how often and when to take a particular drug—that fits with the your habits and lifestyle.

All this works because Moore has employed many of the latest advances in the design of interactive human-computer interfaces, and a few of his own, to make all of this information understandable to mere mortals like me. Using these CollaboRhythm interfaces together, Moore and I have selected an anti-inflammatory drug and dosing schedule for my (imaginary) knee pain, and I now know more about osteoarthritis of the knee and anti-inflammatory drugs than I ever dreamed.

Which is precisely the point: to give patients the kind of information about their conditions, and the deep understanding of what it means, that previously only doctors had. In the CollaboRhythm process, all the information is totally transparent, all the decision-making is collaborative, all the goals are shared. Under this model, the doctor-patient relationship is not unlike the faculty-student relationship at the Media Lab. The doctor is essentially a teacher, the patient is his or her apprentice, and the relationship between the two is a trusted partnership. This seems radical at first, but when Moore gives you a demonstration, it's impossible not to begin to share his deep conviction that the current health care system *dramatically underestimates* the ability of ordinary people—including you—to play an active role in their own health care. Of course, a demo in our *New Media Medicine* workshop is one thing, and actually putting this invention to work in real-world situations is another. That is why, for his master's thesis, Moore decided to tackle a particularly tough problem, and one with huge consequences: medication adherence.

About half of people with chronic diseases like diabetes, high blood pressure, or elevated cholesterol don't take their medications as prescribed. This behavior doesn't just affect them. According to a recent study by the New England Health Care Institute, the health complications that result from patients' not taking medication consistently or properly costs the already financially strapped U.S. health care system a staggering $290 billion annually.

Why do some people with chronic diseases consistently fail to do something as simple as taking their medications—something that has such a dramatic impact on the length and quality of their lives? It's not because doctors aren't warning them about the dire consequences of not taking their meds; they are. One possible explanation is that modern life is so complicated, people simply forget, particularly as they age. But if that were the only reason, then gadgets designed to remind and nudge them to take their medicine, like pill bottles with caps that buzz, or cell phone apps that make your phone ring or vibrate when it's time to take a dose, would do the job. But that hasn't been proven to be the case.

Moore believes there are deeper reasons why people have trouble taking their medications as prescribed. He maintains that patients have a hard time retaining the instructions they're given in the doctor's office and, even when they do remember, they still have little or no understanding of *why* they need to take the medicine in the prescribed way. It's not that they're incapable of understanding; it's just that they haven't been presented the information in the right way. This problem has especially serious consequences for patients whose conditions have no overt symptoms but can turn serious or even deadly if left untreated, like high blood pressure or HIV. After all, why should you bother to take a pill if you are feeling perfectly fine, especially if that pill is causing unpleasant side effects?

In the case of HIV, the situation is especially severe. In the early stages of this disease, there are few if any symptoms, but at some point, with no warning, the virus can overtake the immune system, leaving the body vulnerable to every infection that comes its way. In recent years, several drug combinations, or "cocktails," have proven to be quite successful in halting the proliferation of the virus and preventing it from progressing into full-blown AIDS. But in order for these drugs to be effective, they need to be taken consistently. Missing just one dose a month increases the risk that the virus will mutate into a form that is resistant to the treatment, which usually means switching to a more complicated regimen. Sometimes the virus can become so resistant that it can't be treated at all. Yet studies have shown that as few as 33 percent of people who have been prescribed HIV medication take it consistently or correctly, and previous efforts to improve adherence, including the aforementioned reminder tactics, have had very limited success.

Which is why Moore decided to take a different approach to this problem. He believed that if HIV patients could be given the tools to help them truly *understand* how their medications worked and actually *see* how they are working in their bodies, or not working if they failed to take them as prescribed, then the patients would become more proactive and responsible about taking them on schedule. In order to test the theory with HIV patients, Moore partnered with the Center for HIV and AIDS Care at the Boston Medical Center, where a clinical researcher conducted a month-long medication adherence study.

Each patient began the study with a *planning session,* in which he or she sat together with the clinician and used the CollaboRhythm application described earlier to create a medication schedule tailored to the patient's medical needs and lifestyle. A cell phone was then given to each patient to use when he or she was up and about, and a *Chumby,* to use at home, each of which was loaded with the patient's dosing schedule and an identical suite of applications. A Chumby is a compact, unintimidating device resembling a clock radio with an embedded computer, a touch screen, and a Wi-Fi connection, encased in a colorful leather housing. It provides all the necessary functionality at a fraction of the price of a laptop or a PC, making it accessible to low-income patients, and it doesn't require any special training to use, making it approachable to people who are not computer-literate.

Here's how it works. Let's say you're the patient relaxing at home watching TV. When it comes time to take a pill, the Chumby softly plays a song (you can download a tune of your choice) meant to prompt you to check the *daily medication clock* on the touch screen, which pictorially displays all the pills to be taken right now. Lest you try to ignore it, the song gets progressively louder until you take your eyes off the TV and take each pill, then touch the corresponding icon on the clock face to indicate you have done so. When the icons for all the pills have been pressed, the song turns off.

So far, though, this is just a fancy type of reminder system, which, as I've mentioned, doesn't work. But what happens next is where it gets truly innovative. With a swipe of the Chumby touch screen to the left, you can summon a simple graph that essentially shows the relationship between the pills taken and the predicted concentration of medication in the bloodstream over the course of the current week. The areas indicating acceptable levels of concentration are shaded, meaning that you can easily *see*, in real time, whether there are effective levels of the medication in your bloodstream. Traditionally, doctors and pharmacists have considered information like concentration levels and other such medical data to be too complicated for patients to comprehend. But not Moore. He feels that with the right kinds of tools, patients *can* truly understand the implications of their actions, and as a result, they will be empowered to take control of them.

But this isn't all. Moore's confidence in HIV patients' ability to understand the medical implications of their behavior, and take the appropriate actions, has gone much further. Which is why he programmed the Chumby with an application that, with another swipe of the touch screen to the left, displays an animation simulating what's going on inside deep inside the patient's body, at a microscopic level. Not unlike an animation one might see on educational TV, it shows the CD4 T-cells, which are critical to the cells' signaling pathways of the immune system, literally battling the HIV virus, which penetrates these cells and kills them if they are left unprotected by the medication. If you've been taking the medication properly, the CD4 T-cells are encased in a protective "armor" so that no viruses can attack them. The CD4 T-cells are rich in color and move around the screen fluidly, to show that they are healthy, and the viruses are grayed out and immobile, unable to attack. If you haven't been taking the medications properly, however, the CD4 T-cells are grayed out and don't move, indicating that they are being successfully penetrated by the viruses, which are in full color and moving around in a frenetic, menacing manner. The point is that you can now visually and viscerally understand the impact on your body of failing to take your medications, and you can get back on track before it's too late.

Is putting this information in the hands of the patients really enough to reverse years of bad habits and get them to start consistently taking their medicine? So far, signs point to yes. In Moore's trial with HIV patients, all of the test subjects not only embraced the technology, but three quarters of them achieved optimal adherence at a rate greater than 95 percent.

Since the sample set for this trial was limited and the time frame short, Moore still has a long way to go to before he can prove that his vision for patient empowerment can really work. However, these promising early results have been enough to turn many of his early skeptics in the health care community into believers. He'll soon be rolling out pilots with three major hospital systems in the United States and one in Europe; and a number of Media Lab sponsors—including office furniture supplier Steelcase, insurance provider Humana, P&G, and Samsung—have shown great interest in what Moore's vision for the future of health care could mean for their businesses.

EXERCISING "SHELF" CONTROL

Moore's approach to enabling ordinary people to take control of their own health is based on the conviction that if given access to information about their health, and the deep understanding of what it means, patients can collaborate better with their physicians and play a much more proactive role in their own care. But health is a complicated puzzle, and much of it depends on the behavior that we exhibit in the course of our everyday lives. This is particularly true when it comes to behaviors that can mitigate the effects of disease, or even better, prevent disease in the first place.

Consider the much-discussed problem of weight control. According to the Centers for Disease Control (CDC), two-thirds of all U.S. adults over age twenty are either overweight or obese, which puts them at risk for everything from heart disease to hypertension, diabetes, kidney disease, stroke, and even some forms of cancer. Sadly, children are not exempt from this epidemic, either. Nearly 18 percent of all children between the ages of six and eleven and 17 percent between the ages of twelve and nineteen are over their healthy weights. First Lady Michelle Obama has adopted combating childhood obesity as her major public campaign, called *Let's Move,* and this has done much to raise awareness of the challenge as well as possible solutions.

The fact is that most people are aware that the key to losing weight and keeping it off is to eat healthier and exercise more. But as anyone who has ever attempted to diet knows, it's not that easy. When it comes to controlling our weight, simply knowing what we should do—from every morsel that we put in our mouths to every flight of stairs we take, or don't take—isn't the same as actually doing it on a day-to-day basis.

Then there is the even bigger challenge if we want to actually change our behavior on a consistent, long-term basis. Health and nutrition experts have offered countless strategies for sticking to a diet, selling untold millions of books in the process, but none have been shown to really work over the long haul.

Enter Cory Kidd, a recent PhD recipient from the Media Lab, who has been working for years to invent a tool to help us do just that. At first glance, Kidd, whose background is in computer science and robot design, seems hardly a likely candidate to become the next weight-loss guru. But the idea isn't as farfetched as it seems because Kidd has a vision for how social robots can help people modify their behavior to achieve their long-term health goals. To bring this vision to the world, he decided to start with helping people stick with their diets and founded a startup company called *Intuitive Automata, Inc.* (which he brewed up at MIT's Muddy Charles River Pub while serving as its manager and at the same time pursuing his doctorate at the Media Lab).

"We as human beings are really good at making changes. And we are really terrible at sticking with them. A diet is a perfect example," Kidd often says in his stump talks these days. "When someone starts a new diet, he or she is motivated to stick with it and believes that this time it's going to work." But, he then explains, the average diet is only three and a half weeks long, and about 98 percent of dieters fail in their diets, not because they don't initially lose weight but that those who do almost invariably gain it back, sometimes more than what they lost. Then, he brings out the real star of the show: a white plastic robot just over a foot high, named *Autom*, the first product of Intuitive Automata.

Autom has two stubby feet enabling it to stand on a counter or tabletop; a rectangular body covered mostly by a full-color touch screen that's programmed to log and track food that is consumed, exercise, and weight; and a face whose most striking features are its winsome baby blue eyes. With a camera inconspicuously built into one eye, and face-tracking software that enables her to locate and make eye contact with you wherever you are talking with her, Autom can also move her head up and down in agreement, blink her eyes, and speak to you in a soothing, gentle voice. Her purpose is to be your own personal, friendly weight-loss coach, providing the critical social support to motivate you to make better choices throughout the day, and most important, to stick with those choices over time. Autom provides the extrinsic encouragement that gives people the intrinsic motivation to take control of their weight, and hence, one of the most important factors in their health, not to mention their happiness.

The inspiration for Autom goes back to when Kidd was working on his PhD in Cynthia Breazeal's *Personal Robots* group, where he was her first graduate student. For three years he also spent time working at the Boston Medical Center's Nutrition and Weight Management Center, and while there, he observed the techniques used by professional caregivers to help people lose weight. He noticed that the most effective coaches were those who made upbeat, encouraging comments to enforce consistent positive behavior and offered helpful but subtle nudges such as "Since it's afternoon, you've probably been able to get some exercise in today, right?" when people were doing less well. So he decided to build a social robot that mimicked the motivational styles of these weight-loss coaches as closely as possible—one that could even make small talk about the weather, and occasionally crack jokes, just as a human coach might do.

But could a robot, no matter how sophisticated, or "social," truly have the same effect as a living, breathing weight-loss coach? Could it actually keep a person motivated over time so that he or she could achieve his or her goals? Kidd decided to find out. He first designed and then hand-assembled seventeen research prototypes of Autom in the Personal Robots workshop. In addition to an expressive head and face, each sported a touch screen on its midsection to allow the dieters to easily log and track information daily about what they ate and the exercise they got and to view graphs and charts showing progress against goals. Kidd also programmed Autom with a host of conversational prompts, questions, and "pep talks" that mimicked the best of the professional weight-loss coaches he had studied.

When the Autom prototypes were ready for testing, Kidd lined up forty-five people in the Boston area who had a history of failing to stay on their diets, yet were determined to lose weight. He then introduced Autom into a third of their homes, helping people to set her up in their kitchens, living rooms, studies, and bedrooms. Another third of the dieters were given a touch screen computer, programmed with exactly the same logging and tracking and conversational software as Autom, but without the expressive face and those deep blue eyes that connected with you while she spoke. For completeness of the test, the remaining third of the dieters were given just a paper log to write down and monitor their progress, as that is today's most common diet-tracking tool.

Not surprisingly, about halfway through the six-week trial, most of the people using the paper diary had dropped their diet. And not long thereafter, the people using the computer had also given up. What was amazing was that nearly all of the people who had been given the robot lasted a full six weeks on their diets. In fact, many of them had grown so emotionally attached to the Autom, they had a hard time giving her up when the study was over.

Kidd tells the story of one woman who had outfitted the robot with a Red Sox hat and had named her Maya. She used Maya religiously throughout the test period, and when Kidd went to pick up the robot at the end of the test period, the woman referred to her affectionately by name. She even asked Kidd if she could talk to Maya one more time before he took her, and she came out to the car to wave Maya a final emotional goodbye as Kidd pulled away. And this person wasn't an anomaly, as nearly everyone who had a robot in their home had named it.

This response was exactly what Kidd was hoping for. Kidd wants Autom to be much more than just a personal computer on which you can record what you eat and how much you exercise. What makes Autom so much more effective than a simple computer, as Kidd's research showed, is the emotional, humanlike bond she forges. The idea is that if the dieter develops an emotional connection to Autom, he or she will be more likely to talk to her every day about how their diet is going, and as a result the dieter will be motivated to make more of the "rational" choices that lead to weight loss. In other words, by making you feel truly accountable for your decisions, Autom helps you to make better ones.

But soon you'll actually be able to talk to her naturally about what you ate and how much you exercised that day; and in future versions she'll be able to learn more and more about you and your daily lifestyle. For example, it's easy to imagine Autom automatically collecting information throughout the day, transmitted through your smart phone, from sensors on your body as well as sensors in the furniture, walls, and floors of your home. This would make possible a friendly collusion between Autom and Linda, the anthropomorphic conversational agent on your TV. Don't you think you would be more likely to pass up that candy bar or resist the temptation to skip your workout if you knew that your health coach robot was dutifully reporting back to your doctor through your health liaison avatar?

Kidd has a lot in common with Sajid Sadi, a current PhD student in Pattie Maes's *Fluid Interfaces* group. Like Kidd, Sadi believes passionately that many of the health problems endemic in our society come back to our behavior. Whether it's eating too much, exercising too little, forgetting our medicine—each behavior is controlled by habits over which we have much less control than we'd like. Yet most approaches to these problems today don't really try to change our habits; they merely try to suppress them, by limiting our options. Think about those popular weight-loss programs that urge you to eat prepackaged foods and preplanned menus. The idea behind these is that if you can't choose what you eat, you can't choose badly. But because the regimen is imposed and you don't get a chance to practice making the right choices, the bad habits survive—which is why studies show time and again that people gain back the weight as soon as the imposed regimen stops.

But before we can attack our bad habits, Sadi maintains, we must first learn to understand them. For example, how do we make real-time decisions about the food that we eat, as we are literally in the process of eating it? It is easy to make a resolution not to eat too much, but when we are eating, we don't always stop to consider how we're doing. The decision to have one more bite is often based on social signals, rather than the taste of the food and how full we feel. Sadi calls these *micro-decisions,* and he is exploring how technology can help arm us with the information to make them more rationally, and less automatically, in a way that supports our *macrogoals.*

Sadi is working on a number of prototypes, which he calls *Reflect-Ons,* of tools that help people be more aware of these microdecisions *as* they are making them and reflect on their implications. One is a fork equipped with a tiny motion sensor on the handle that detects when the user takes a bite of food. While it can't currently tell how much and what exactly you are eating, it can tell how *fast* you're eating—important information given that many of us overeat because we're eating so quickly that our digestive system doesn't have time to signal our brain that we're full. The fork also measures the time between bites and lets us know we're eating too quickly by flashing a light or gently vibrating. While this isn't necessarily a "stop eating" signal, it is a reminder to eat more slowly. It also encourages us to be more conscious about our eating in general and to think about our larger health goals *while* we're eating the meal, rather than making us feel guilty about what we've eaten *after* we finish the meal (as many diet tools do).

As with the other devices I've described, the point is not to make people's choices for them but to give them precisely the right information, at precisely the right place and time, and to encourage them to make the right choices on their own. Only then can they truly break free of their bad habits and replace them with better, healthier ones that will stick over the long haul.

THE MONEY CURE

If you spend a few hours chatting with Charlie DeTar, a doctoral student in Chris Schmandt's *Speech + Mobility* group at the Media Lab, you'll likely come away thinking—as I did—that health care isn't the only area where we might be able to encourage better, more constructive behavior by arming people with more information about the consequences of their everyday choices and providing them with the tools to help them make better ones. DeTar believes that impulsive spending is another social epidemic, like obesity, that often results from "irrational" short-term decisions that may be pleasurable in the moment but make us unhappy in the long run.

DeTar's 2009 master's thesis project, called *Merry Miser,* was certainly well timed, coming as it did in the midst of the global financial crisis. By now it's widely known that out-of-control consumer spending was one of the many factors contributing to the precipitous meltdown. But DeTar maintains that this storm was brewing for the past two decades, as advances in information and communication technologies have enabled the marketing and advertising industries—or what he calls the *persuasive media*—to become more and more effective in influencing consumers' behavior.

For example, the explosive growth of the mobile phone industry and advances in the mobile phone platform such as location-tracking software have given the purveyors of persuasive media new ways to influence people to buy more. Today, using sophisticated real-time data-mining algorithms, the persuasive media can "respond in real time, personalize its message, and interact in much richer and deeper ways than older media like TV, radio, or print advertisements." He cites systems that send advertisements, promotions, and coupons that are personalized according to a person's unique habits, spending profiles, and current location, straight to his or her mobile phone. To that I would add new technologies like *near field communication* that make it faster and easier to make purchases on the spot using your smart phone, whenever the impulse hits.

Yet, DeTar believes that the very same advances that have made us sitting ducks for advertisers and turned our lives into nonstop shopping sprees can actually be used to help us *curb* our spending and take control of our finances. Just as is the case with health care, too much information is currently part of the problem, but with the right tools, it can instead become part of the solution.

That's exactly the point of Merry Miser, a location-based service for smart phones that, instead of giving companies more information about our spending behavior patterns and habits, puts that information *back* in our hands, then helps us use it to spend less, and save more. Merry Miser is effectively an *intervention*—a tool to help you be more aware of your habits and understand the long-term consequences of your decisions *before you make them,* like at the moment just before you take that first step into a shop, restaurant, casino, or anywhere it knows, based on your past behavior, you are likely to be tempted to spend money. A true antidote to impulse shopping, "the goal is to make you more aware of your desires and impulses at the moment and how you're going to view your purchases in the long term," DeTar explains.

Here's how it works. Let's say that you frequently shop at a local clothing store. Every time you walk near the store, your Merry Miser, which knows exactly where you are from the GPS in the smart phone, will buzz and send you a text message reminding you how much money you have spent in that shop in the past and the amount of your average purchase. It will also remind you that when you were asked a few days after the purchases how you *felt* about them, they had quickly lost their luster for you. In fact, Merry Miser will remind you that you gave the overall shopping experience at that store a lukewarm rating, and when asked how you felt about your purchase a full month later, you admitted it was still hanging somewhere in the back of your closet, with the tags still on. Then, in case you didn't get the picture, Merry Miser will inform you that your "happiness" rating for shopping at this store is LOW. If that isn't enough to make you resist that sweater in the window display, Merry Miser will show you a simple visual image depicting the current state of your finances relative to your goals. And finally, just to be sure you remember the value of being thrifty, it will show you a poignant and motivational quote, like this one from Bertrand Russell: "To be without some of the things you want is an indispensable part of happiness."

As you can see, Merry Miser is less about managing your money and more about managing *your information* about your impulses, desires, and habits, and the emotions behind them, to help you make more reasoned, rational spending decisions. "There is no moment when Merry Miser tells you 'don't spend,'" DeTar explains. "The information it presents is relatively neutral in respect to whether or not you should make the purchase. The purpose is to make you more aware of exactly where your money is going and how this is contributing to your well-being. It helps you to make a better decision." Again, we see how technology can be used to right the imbalance between what other people—in this case, businesses—know about us and what we know about ourselves, and it can put the power back in our hands.

Merry Miser is currently in its prototype stage, being tested by a handful of subjects. DeTar acknowledges that in order for a service like Merry Miser to take off, banks will need to participate by making some changes in how they record their transactions, making certain information more public. Given how much money the banks make from consumers' overspending today, he is sure that this will take some time. But I believe that eventually, technologies like Merry Miser will convince banks and other consumer businesses that in the long run they will make more money by helping people spend their money wisely. Everyone will be better off, and perhaps we will avoid a recurrence of the irrational behavior that helped bring us all to the brink of disaster just a few years ago.

FURTHER REFLECTIONS

Our health care and financial systems are in the same sickbed, both suffering from the failure to recognize the power of ordinary people to play a more active role on their own behalf. To truly heal these systems, it is not enough to just reform their *institutions,* as is the focus of public policy these days. Instead, we need to radically change the role of *individuals* within the systems, eliminating the age-old asymmetry between themselves and the "high priests" of medicine and money. This is the only sustainable approach.

When I first told people of my plans to start an initiative

at the Media Lab to revolutionize health care, they looked at me like I was crazy. Why would the Media Lab be a place to address the complex challenges of the medical system? It was simple, I explained. The Media Lab brings radically new thinkers, and radically new ways of thinking, to existing problems. That's precisely what the healthcare system needs.

Today, several years later, I like to lead visitors on what I call a "health of the future" tour of the Media Lab. We start in the New Media Medicine "living room" with the demonstration of CollaboRhythm you read about earlier in this chapter. We then walk a few feet across the East Lab to stand in front of what appears, at first glance, to be a simple wall mirror. But after looking at it for a few seconds people are usually surprised to see a digital display of their heart rates appear below the reflected image of their faces. This is *Cardiocam,* the brainchild of Ming-Zher Poh and Daniel McDuff of the Affective Computing group. They mounted a standard video camera pointed at the viewer's face from behind a one-way mirror and developed an algorithm that analyzes the light reflected off his or her skin to deduce the subtle volumetric changes in blood vessels that occur every time the heart beats. It will soon also display blood oxygen level and blood pressure. It's the forerunner of a whole new class of biometric sensors that will be hidden in clothes, smart phones, household products, furniture, bikes, cars, and more, providing individuals the information they need to seize control of their health.

We then proceed up the stairs to the *Camera Culture* workshop, where post-doc Ankit Mohan and visiting student Vitor Pamplona await us. Pamplona begins his demo with a mobile phone in one hand and a small white plastic eyepiece unit in the other. He slips the eyepiece unit over the phone's display and hands the combination to one of the visitors, instructing him or her to look into the eyepiece and align the red and green lines that come into view by pressing buttons on the phone keypad. This is repeated eight times, taking less than two minutes, and at the end software in the phone computes the person's refractive errors and determines his or her eyeglass prescription. Pamplona and Mohan's invention is called *NETRA,* which means "eye" in Sanskrit and is also short for "near eye tool for refractive assessment." It is essentially a thermometer for vision.

NETRA grew out of the scientific vision of professor Ramesh Raskar, a pioneer in the emerging field of computational photography and a recent addition to the Media Lab faculty. As a youth in India he devised a method for attaching a projector to a camera to display family photos since printing was prohibitively expensive. Today, as head of the *Camera Culture* group, he aspires to literally reinvent the camera in the digital age. His group is developing radical new forms of visual information capture and analysis (such as a camera that takes pictures "around a corner") that could transform entertainment, worker safety, and health for billions of people. To that end, Raskar is working on getting the manufacturing cost of NETRA down to a few dollars apiece so that it can be distributed widely. Since uncorrected refractive errors are the second leading cause of blindness in the world, and the cause of debilitating stress and headaches for many children in developing regions, this simple device could help prevent untold human suffering, as well as huge economic loss. Raskar is also directing his group to develop a mobile phone clip-on eyepiece for self-assessments for cataracts, the number-one cause of blindness in the world.

We exit the Camera Culture workshop and proceed to look at a variety of other inventions that put ordinary people in command of their own health, some of which you have read about in these pages, and others still in the earliest stages of prototyping. By the time we finish, my visitors have a new appreciation for what can be accomplished when you bring radically new thinkers, and radically new ways of thinking, to the challenges of health care.

CHAPTER 8

I Am a Creator

"My Eagle Song," a synthesized voice announces. The conductor, Dan Ellsey, who is also the composer, is seated center stage. He waits a few seconds; then with a dramatic swing of his head upward and back, he cues in the first note. An electronic harpsichord begins to play a short melodic tune, as Ellsey swings and lurches, his head cutting lines through the air, landing note by note in step with the naked melody. The digital instruments follow his cues, as he wrestles the timbre—sometimes softer, sometimes sharper, sometimes more shrill—from each note.

I am seated toward the back of the auditorium, which is filled with the eclectic mix of the artists, entrepreneurs, writers, media moguls, movie stars, academics, and thinkers who have come to TED 2008 in Monterey, California, to explore "the power of ideas to change attitudes, lives, and ultimately, the world." Over the past few days they have seen all kinds of amazing things on stage, but from the instant that Ellsey cues the first note, I observe they are riveted in their seats and engaged emotionally in a way they hadn't been for any of the other performances or speeches.

Meanwhile, on stage, Ellsey is still in a torrent of motion, throwing his head back from time to time to cue one particular note or another. His hands open and close, clawing in front of him as his head dances in time with the music. I can't tell you exactly what is happening in the turbulent eddies of his body, but one thing is clear: The audience succumbs to his musical interpretation.

Suddenly, Ellsey slows his pace. The whole piece has been building up to this moment, this underlying heart of the work. Larger, louder, Ellsey's movements are increasingly emphatic. With each triumphant dash of his head, he cues a new instrument, as the auditorium fills with the sound of a mandolin, a banjo, and something akin to a Chinese lute.

Ellsey now flashes a broad smile as he prepares us for the final buildup. He gives us three last notes, each one a pitch higher than the last. His eyes shoot upward with each note, the last of which is held for several seconds before he gives his digital orchestra that final cue, and the room falls silent. That is until the crowd jumps to its feet and erupts with applause.

As the audience filtered out for the break, I stayed put in my seat, emotionally drained, just observing the scene. I remember my attention being drawn to a fellow who also remained seated, about five rows ahead of me, and who appeared to be so moved he was wiping away a tear. But I knew it wasn't just the music itself that the audience had found so moving. Just before the performance, we had been told the story of how Dan Ellsey had always loved music, but for the first three decades of his life, he had no choice but to sit back and watch other people perform it. Born with cerebral palsy, a degenerative disease that causes a loss of motor control, Ellsey, who is bound to a wheelchair, is unable to walk or talk, let alone hum a tune, pick up an instrument, or use a pen to jot down the notes that are constantly playing inside his head.

For today's performance, Ellsey, a resident of Tewksbury State Hospital, about twenty miles northwest of Boston, is 3,000 miles from his home. This is the first time in his life he has ventured outside the border of his native Massachusetts. A few months after TED 2008, I had the opportunity to visit the Tewksbury State Hospital for the first time, and I experienced for myself the magic that had transformed Ellsey into a composer and conductor. I was the guest of honor at a celebration marking Ellsey's memorable TED performance, and the hospital staff planned to thank the Media Lab for making it possible. As I proceeded in my car up the long drive, I couldn't help thinking that the main building, a Queen Anne–style dark brick structure with a copper-clad clock tower visible through a wrought-iron gate, could easily serve as a movie set for a nineteenth-century asylum. I later learned that in fact it once was one.

Inside I found something quite different than the imposing exterior would predict: a vibrant and enthusiastic hospital staff eager to show me the innovative programs they had developed over the past few years using visual, musical, and performing arts as forms of therapy. These programs centered around helping patients—many of whom couldn't speak, write, move their facial muscles, or express themselves in any traditional fashion—achieve a level of emotional and creative expression previously believed impossible. As I walked through the hospital wards encountering patients with physical and psychological conditions more debilitating even than Ellsey's, the magnitude of that challenge faced by this intrepid staff suddenly overwhelmed me.

The last stop on the tour was the music room where Ellsey and a half-dozen other residents were anxiously awaiting us. To my great surprise, rather than the keyboards and percussion instruments I expected, there were several banks of desktop computers and a variety of high-tech gizmos and digital devices not unlike the ones you would find in Tod Machover's *Opera of the Future* workshop at the Media Lab. Then Ellsey, seated in his wheelchair and wearing an infrared tracker on a headband, proceeded to give me a five-minute demo of the technology he used to create and perform "My Eagle Song," just the way a Media Lab student would do it in his or her workshop in Cambridge. At the conclusion of his performance, the staff and I broke out into both tears and applause, just as the audience had the month before in Monterey. Later that afternoon, as I proceeded down the long drive on my way back to Cambridge, the main asylum building in my rearview mirror, I called my wife to tell her that I had just experienced the most rewarding day of my professional life.

I AM A MUSICIAN

The revolutionary technologies I saw that day at Tewksbury State Hospital—the ones that turned Dan Ellsey into a noted composer and performer of digital music—were actually the brainchild of someone you read about in an earlier chapter: Media Lab Professor Tod Machover.

A cello prodigy who had a passion for technology from a very young age, Machover was raised by a Julliard-trained pianist-turned–music-teacher mother and a computer graphics–pioneer father. Machover recalls how his mother, Wilma, who believed music was not just something written by dead people and performed by classically trained virtuosos, would after every music lesson tell her pupils to go through the house and find objects that could make interesting and beautiful sounds. "We would bang on tabletops and lamps and anything else that we could get our hands on. We went a little crazy, but learned that music emerges from the world around us and that anyone can make it," Machover remembers.

In the early 1970s, when he was a high school student at the Fieldston School in Riverdale, New York, Machover's own music teacher arranged for him to meet a famous Fieldston alum who had been a gifted musician but had ended up as a computer scientist. It turned out to be Marvin Minsky, a cofounder of the *Artificial Intelligence* laboratory at MIT. Machover found a kindred spirit in Minsky—someone who was equally passionate about both music and technology and who thought about music in an untraditional way. The fact that one of the most original thinkers about music was someone famous for something completely different left a lasting impression on Machover.

After earning his bachelor's and master's degrees in music composition from the Juilliard School of Music, Machover spent seven years in the early 1980s in Paris at Pierre Boulez's IRCAM (Institut de Recherche et Coordination Acoustique Musique), a brand-new facility at the Centre Pompidou where artists and technologists intermingled to study the science of music and sound, working to create avant-garde compositions of classical music. Here he got to experiment with the range of new digital technologies—including the world's first real-time digital synthesizer—that made it possible to play with music in new ways, to actually *touch* and *feel* music. "I wanted to throw it around a room and bounce it off the walls, the way I did in my mother's living room," explains Machover.

There was no academic institution in the world other than IRCAM that Machover felt was suited for his work—that is, until he got a phone call in 1985 inviting him to head the new MIT Media Lab's music program. Imagining collaborating with the likes of Marvin Minsky, he leapt at the offer. He started the *Opera of the Future* group, and they immediately began inventing *hyperinstruments,* which you'll remember from Chapter 4 are traditional musical instruments such as cellos, violins, pianos, and percussion augmented with digital technology such as motion sensors and transmitters that push the limits of virtuosic performers' expressive powers. One such hyperinstrument was the hypercello developed for Yo-Yo Ma.

But in the mid-1990s, Machover decided to take the Opera of the Future group in an entirely new direction. He wondered, what if, instead of building hyperinstruments to help the world's greatest performers play even better, they explored what these same technologies could do for the rest of us—people, like me, with little or no musical training and a complete lack of any natural musical talent. So, perhaps thinking of his own young daughters, then just beginning their musical journey, he decided to focus his group on inventing new ways to enable every child on the planet—even those who had never sung a note, played a chord, or touched the keys of a piano—to experience the sheer joy of making music.

There was something else on Machover's mind that steered him in this direction. At the time, there was a lot of buzz in the media about the so-called *Mozart Effect*, a term that grew out of the results of a study by neuroscientists Gordon Shaw and Francis Rauscher of the University of California, Irvine, that had appeared a few years earlier in the journal *Nature*. They had found that college students who listened to a Mozart sonata for ten minutes before taking an exam scored higher on the questions designed to evaluate their temporal and spatial reasoning, which was thought to indicate better math and science skills. Even though a closer look at the research revealed that the bump in IQ was only temporary, the media (and in turn the public) inexplicably took this finding to mean that children who listened to Mozart would grow up to be smarter and more successful in school and life. Soon, parents everywhere were piping Mozart sonatas into their children's cribs; some overeager expectant mothers even started walking around with headphones clamped onto their stomachs in the hopes that the music would reach their unborn children in the womb.

While Machover agreed that music does have positive effects on the brain, he didn't believe these benefits could be attained quite so passively, just by listening. He knew that composing and performing music are two of the most stimulating activities that humans can partake in because they engage more parts of the brain than any other activity and create all kinds of connections between different parts of the brain that wouldn't exist otherwise. He suspected that music could have truly positive and perhaps long-lasting effects on the brain, particularly for young children, provided they were actively engaged in making it, rather than simply listening to it.

But there were a number of barriers to getting all children involved in making music. First, the cost of traditional instruments and lessons is prohibitive for many people in most parts of the world. But even for those who do have the means, truly mastering an instrument requires an incredible amount of time, patience, and dedication, and many children are either too intimidated to try or they quickly get frustrated and quit.

Machover and his Opera of the Future group set out to topple these barriers by developing a second type of hyperinstrument, first for the general public in their celebrated *Brain Opera* project (premiered at New York's Lincoln Center and now installed permanently in Vienna's House of Music), and then especially for children. These new hyperinstruments didn't look like or cost as much as a traditional instrument and didn't require years of practice in order to play. The first ones, called *Beatbugs*, were small handheld "creatures" shaped like bugs, and they allowed children to create well-constructed, drumlike rhythms simply by tapping on their backs, and then modifying the rhythms by bending the sensor "antennas." A wireless network between Beatbugs allowed children to share their rhythms with one another, creating more complex compositions together. Next came *Music Shapers*, which were like Beatbugs except that they were covered in soft, conductive material, and children could "play" them by touching, squeezing, twisting, or stretching.

Then, Machover and his students launched an ambitious three-year project that they called the *Toy Symphony*. The idea was to bring groups of children together with some of the world's greatest performers, like violinist Joshua Bell and conductor Kent Nagano, and with symphony orchestras, such as the Deutsche Symphonie Berlin and the BBC Scottish Symphony, for weeklong workshops at various locations across the globe. Each one culminated in a grand final concert in each city's symphony hall, where together, they would all perform specially commissioned pieces by Machover and others, as well as new compositions created by the local children, who also played their Beatbugs, Music Shapers, and other music toys.   ·

But perhaps the most revolutionary technology introduced in the Toy Symphony wasn't an instrument but rather software for composing music called *Hyperscore*. Developed by Machover's students Mary Far-bood and Egon Pasztor, Hyperscore is designed around an intuitive and expressive interface that lets you create a musical score by essentially "painting" it onto a "canvas" using a mouse or other pointing device. Getting started is fast and easy. You begin by placing eggplant-shaped "notes" in boxes to create melodies, chords, and rhythms, then assign each of these "motives" a distinct color. Finally you draw lines of vary-ing lengths, shapes, colors, and texture onto the canvas, and all of this is automatically converted into musical notes. What you are actually doing is creating a string of thematic melodies. Then all you have to do is click the play button, and Hyperscore will transform your graphical creation into music for you to hear. At this point the score might sound like a bit of a mess—but Hyperscore automatically converts the various layers of melodies into beautiful harmonies by letting you draw with a "harmony line" that magically shapes the chords and progressions into meaningful music. All the rules of Western music are right there in the software, called up when needed to make the sonic elements fall into place.

Using Hyperscore, children can dive in and compose a sophisticated song in just minutes. By experimenting with many different kinds of melodies and sounds, they can learn volumes about what makes a com-position great—its structure, themes, climaxes, and contrasts—without getting bogged down in the technicalities of music composition that take years of experience to fully understand. Seeing a group of children who have never held an instrument in their hands compose an entire musical score in just a week, as they did for the Toy Symphony, was all the proof Machover needed that we all have an "inner Mozart." All we need is the right tool to unleash it.

Then, as often occurs at the Media Lab, another path for this technology serendipitously presented itself. In 2004, after one of the Toy Symphony's final performances, a member of the Massachusetts Cultural Council approached Machover with an intriguing question: Could the Toy Symphony technologies be used as a form of music therapy for residents of Tewksbury State Hospital, all of whom have either severe neurological or psychological disorders? The staff at the hospital had previously developed a number of successful programs in which residents were using music, the visual arts, or theater to communicate and express themselves in ways that their disabilities had made very difficult, or even impossible, to achieve. The staff was now eager to experiment with new technologies to help them take this approach to the next level. This resonated with Machover, who had already been considering how his music technology could help those who have experienced strokes, various neurological problems, and even common depression. He was particularly fascinated by reports that described how music is one of the last things to which later-stage Alzheimer's disease patients will respond.

He accepted the Tewksbury challenge. It was a perfect project for one of his newest graduate students, Adam Boulanger, who just like him, had grown up in a household in which experimenting with both music and technology was encouraged. Boulanger remembers attending a computer music festival at the age of eight with his father, an early computer musician, during which a performer wired up a chair and then threw it around an enclosed space. Every time the chair would collide with a different surface, it would produce a variety of rich, textured sounds. "I was pretty impressed by that," the soft-spoken doctoral student recalls with a grin. But Boulanger wanted to do more than build technology that made cool sounds, so he went on to earn a degree in music and a diploma in premed, and then he joined Machover's group to help bring the power of music to people with disabilities, like the patients at Tewksbury.

Boulanger believed that Hyperscore, although developed originally for the Toy Symphony, would be the perfect vehicle for this work. So together with Machover and a group of volunteers from his alma mater, the Berklee College of Music, Boulanger set up a bank of computers with Hyperscore and began teaching the Tewksbury residents how to use it, similar to what had been done with the children for the Toy Symphony. The early results exceeded even Boulanger's and Machover's expectations: Every one of the residents who tried out Hyperscore, even those with the most debilitating of conditions, was able to successfully complete compositions. Even more amazing, though, were the improvements the doctors and hospital staff began noticing in these patients' mental well-being: They seemed to be developing better coping strategies, showing signs of higher self-esteem, and engaging in fewer negative behaviors, including suicide attempts. Many openly credited Hyperscore for the changes in their attitudes and perspectives. "There was new hope," one resident recalled, as she moved out of the lockdown unit and back into the community.

However, there was one resident in particular who exuded far more enthusiasm about the Hyperscore project than any of the others. His name was Dan Ellsey. Because his cerebral palsy made it difficult to control his motor movements, Ellsey couldn't use the normal handheld controllers that interact with Hyperscore. Nonetheless, Boulanger and Machover were determined to devise some means for Ellsey to overcome this barrier so that he could participate in the program.

For the first few weeks, Boulanger just observed as Ellsey listened to his favorite music. He noticed that even though Ellsey couldn't move his arms or legs, or speak without the assistance of a voice synthesizer, he did have some control over his head movements, and he could express his emotions through his gestures while listening to the music. He also noticed that Ellsey was capable of an array of facial expressions that spoke volumes about what he was feeling inside—including a broad grin when he was happy. Despite his extreme physical limitations, Ellsey was markedly expressive. That's when it struck Boulanger that Ellsey, who was clearly passionate about music, might have profound musical talent lurking beneath the surface—he just needed some tools to help him unlock it.

Boulanger started by designing an infrared controller that Ellsey wore on his forehead that enabled him to compose with the Hyperscore by moving his head to point the beam at lines and colors on the screen, and then arranging them in such a way as to create simple melodies. Once he mastered the infrared controller, it became obvious that Ellsey knew exactly what he wanted to compose, and once he started a piece, there was no stopping him. Slowly but surely, he entered note after note, line after line, his enthusiasm and emotion building along with the piece until the final chord came. When each composition was complete, he could barely stay in his seat.

The crowning achievement of the Hyperscore workshops at Tewksbury was a public concert by the Lowell Philharmonic to celebrate the 150th anniversary of the hospital. The orchestra performed four Hyperscore pieces that had been composed by the patients and transcribed for a full orchestra. Ellsey's piece was the finale and brought the house down.

After the initial Hyperscore project at Tewksbury, Boulanger and Machover discussed next steps. What if Ellsey could actually *conduct* his own Hyperscore compositions, moving from the back row of the theater to the stage? Could he possibly command the tempo and phrasing of the electronic music, imposing his own interpretation with movements of his head alone, just as a conductor of an orchestra does with arm gestures and a baton?

They knew this was incredibly ambitious, but worth a try. Boulanger worked with Ellsey for six months in the music workshop at Tewksbury to develop an ingenious performance system tailored especially to him. Boulanger began by developing a remote sensor–based system for precisely tracking the position of Ellsey's head in space, an updated version of what had been used for Penn and Teller's Spirit Chair more than a decade before. Day after day, he listened to the music and recorded Ellsey's head movements in order to figure out what specific movements he intended in terms of shaping the music—for example, strong linear movements created cues and changed orchestration, whereas distinctly curved movements of various sizes and speeds controlled dynamics and phrasing. As he went along, Boulanger literally coded software algorithms over Ellsey's shoulder to map his reproducible gestures to meaningful musical controls in Hyperscore. Something like a conductor, something like a solo violinist, Ellsey was able to use the final performance system to shape the expression of his compositions with his head as he sat in his wheelchair.

Finally, in the spring of 2007, the system was ready, and Machover decided to feature a live performance by Ellsey as the grand finale of the Media Lab's upcoming Human 2.0 Symposium. Ellsey wrote a new composition for this occasion, "My Eagle Song," and Boulanger recalls, "We would rehearse the piece as you would with any artist. Tod would come in, and the three of us would hash out what makes a killer performance, refining the composition along the way."

Ellsey's performance at the Kresge Auditorium (just like the subsequent one at the TED 2008 that I described at the beginning of this chapter) culminated in a lengthy standing ovation and tears everywhere. But this wasn't just a moving performance for the benefit of the audience; it was literally life-changing for Ellsey. A few months later, when a *Boston Globe* reporter asked him what he wanted to tell people through his music, Ellsey, typing away with a head-operated laser pointer on his computer, and flashing a broad smile, replied, "I am a musician."

I AM AN ENGINEER

A father and his two young daughters are standing in front of the *Living Wall,* a surface covered from floor to ceiling with the brightly colored, interactive wallpaper designed by Professor Leah Buechley's *High–Low Tech* group. When one of the girls glides her small hand across the surface, a 3-D flower embedded in the wall magically lights up—as do the faces of the children. With its bold fuchsia and green floral print on a burnt orange background, the Living Wall looks like it should be on display in a design showroom, rather than on the fourth floor of the new Media Lab. As you might have gathered, though, this is no ordinary wallpaper. Kind of like a giant home remote controller, it could be used to interface with electronic systems—everything from lights to televisions to security systems or even coffee makers, all while being part of the décor. The interactive wallpaper is such a big hit with this family that the father hands his camera to a passerby and asks him to take a picture of them standing in front of it. Nearby, Buechley is beaming.

She wants to spread the word that engineering doesn't have to be an intimidating science used only for building things like cars, computers, and robots. Rather, it can play an important role in the fabric of our everyday lives. Just as Tod Machover believes that each of us possesses an inner musician waiting to be unleashed, Buechley believes that we all have an inner engineer and inventor also waiting to be freed; we just don't have the right tools for doing it yet. She is on a quest to combat what she calls the "cultural lopsidedness of the hard sciences," in particular the gender gap, by getting more females, like those two young girls admiring the Living Wall, interested and active in the world of engineering and inventing. In the long run, she hopes to broaden the very meaning of engineering in our modern society, making it something in which everyone can participate and enjoy, just as she does.

Brought up on a farm in the small village of El Valle, which is an hour north of Santa Fe, New Mexico, she recalls having fun watching her father solve problems on the farm through pure resourcefulness. The effect was that she grew up with the conviction that the best way to contribute to society is by being creative and making things. She was a self-professed "math geek" in high school, and she majored in physics in college. But her real passion was for art, design, dance, and theater, so when she graduated, she worked in the art and design world in New York City for a while.

But she missed technology, and she was also becoming concerned about how she would make a living. After a few years she left the New York City art lifestyle and went on to obtain her PhD in computer science. She did manage, however, to combine her interests in both technology *and* art, becoming a pioneer in the new field of *electronic textiles (e-textiles)*, which essentially are fabrics with electronics of various types embedded or sewn into them. One of her better-known inventions using e-textiles was called the *LED Tank Top*. Slip it on, and a grid of hundreds of LED light bulbs, driven by a microcontroller stitched into the fabric with conductive thread, produces a low-resolution display of John Horton Conway's "A Game of Life," a mesmerizing visual simulation of the biological cell cycle.

But as fun and rewarding as this was for her, building and designing her own e-textile inventions was only the first step for Buechley. She wanted to empower everyone—particularly those intimidated by engineering and the hard sciences—to tap into his or her own creativity and make his or her own inventions. For this she developed the *LilyPad Arduino toolkit*, a set containing various electronic parts, like flashing lights, sound devices, and sensors that can easily be sewn into clothing or fabric using the provided conductive thread. You can use the Lilypad to design and construct your own interactive clothing, like a t-shirt that can measure and display your heartbeat or a skirt that changes colors depending on the speed or direction you're walking. You can even build electronics into upholstery to create smart furniture, like a sofa that plays a song when someone sits on it.

Before LilyPad, only a small number of people in the world designed and constructed e-textiles, and they were mostly professional researchers, engineers, and serious hobbyists. Buechley aimed to change all that. By incorporating technology into an activity that tended to be primarily female—sewing and clothing design—she hoped to be able to get more women involved in the world of engineering and inventing.

Since its introduction in 2007, the LilyPad Arduino has been purchased by tens of thousands of people throughout the world, quite a large number for a traditionally niche product of this kind. But what really gratified Buechley was the fact that about 65 percent of the people building LilyPad projects were female—an extremely high number considering that only about 2 percent of the projects built with typical hobbyist engineering toolkits are built by women.

Buechley is particularly proud of the many stories that users have shared about their experiences and inventions. A college student in California credited the LilyPad with giving her the "courage" to jump into hardware development and create a soft video game controller out of a teddy bear. "Before I started this project, I had absolutely no experience with electronics of any kind. I *still* can't solder to save my life, but it doesn't matter because I can sew."

Then there was the costume designer in Berlin who had an idea that everyone at her theater first told her was ridiculous, if not impossible: She wanted to embed smart blinking lights into the costumes for an upcoming production. She had absolutely no technical experience, but had read about Buechley's LilyPads, and she managed to hunt down a set at a leading electronics shop in Berlin and connect with a community of engineers and hobbyists who helped her design, build, and program the pieces required to implement her artistic vision. She did all the sewing, and she transferred the costumes to the lighting department to change the batteries and proof the circuits for every show. Since that production, she carries a LilyPad Deluxe toolkit wherever she goes, and today she incorporates electronics into all her costumes and set designs. The latest was a coat that appeared to be on fire and spewing sparks. She said that working with electronics in this way "gave her credibility in the largely male-dominated technical theater scene."

Another strong sign of the vitality of the LilyPad community is that it has attracted a number of professional third-party developers who are creating extensions and derivative toolkits using the creative commons license that gives them access to Buechley's schematics and files. As just one example, a team at the Hong Kong Polytechnic Institute has developed a version that enables people to build e-textiles without even sewing. If enough of these third parties jump on board, one day LilyPad, or its successors, may reach beyond niche product status to be broadly adopted by millions. After all, the personal computer was first sold as a do-it-yourself kit to a small band of enthusiastic hobbyists not unlike the LilyPadders.

Buechley's dream, however, actually transcends bringing out the inner engineer in all of us. Since technology has become such an integral part of all of our lives, she thinks everyone should be able to really *understand* technology, and she believes the only way to do that is to enable them to create it, rather than just blindly consume it. Ultimately she would like to see everybody "fall in love with technology," just like she did as a young girl.

I AM A PROGRAMMER

Imagine a website with over a million and a half interactive games, videos, podcasts, articles, and stories, with no monthly subscription fee, no annoying advertising pop-up ads, and no cross-promotion with affiliate websites or any other revenue-generating schemes. Yet this website would have an astounding variety of new interactive media appear daily by the thousands, from video games to music videos, from news broadcasts to e-zines, from social commentary to political satire, from science simulations to science fiction serials, from silly videos of dancing animals to serious tutorials on every imaginable topic.

Want to view the site in Russian, or Cantonese? No problem. At the click of a mouse, you can view the entire site in any one of over forty languages. Want access to the actual software code that drives a game or interactive story that you really like? Just a few clicks, and the code appears in your browser, where you can reuse it and remix it into your own program as you please, without ever having to sign a waiver or see an intellectual property agreement. Would you like to get in touch with the author of a video or podcast or animation that you enjoyed and propose collaborating together on a new project? You could be in business with a team scattered throughout the world in a matter of minutes and have the first products posted in hours.

Does this sound too good to be true? It's not. Just visit http://scratch.mit.edu, and you will find *Scratch,* the brainchild of the Media Lab's Professor Mitchel Resnick and his *Lifelong Kindergarten* group. The Scratch site has been dubbed "the YouTube of interactive media." What may really surprise you is that the median age of content producers for Scratch is twelve.

Scratch was conceived seven years ago when Resnick decided to do for the art of computer programming, a creative skill that until recently was accessible to only a few, what Tod Machover's Opera of the Future group is doing for music making and what Leah Buechley's High-Low Tech group is doing for engineering. In Resnick's mind, empowering children to be actively engaged in programming was the next logical step in his quest to use digital technologies to transform education. "The twenty-first century requires a whole new approach," notes Resnick. "It's no longer good enough to go to school to learn a fixed set of ideas and a fixed set of facts and a fixed set of skills. You need to learn to think creatively, reason systematically, and work collaboratively." His approach is to develop the tools to enable everyone to "learn playfully" throughout the course of his or her entire life by designing, creating, experimenting, and exploring—just as children do in kindergarten.

Ever since its beginnings more than fifteen years ago, the Lifelong Kindergarten group has been creating and bringing to children everywhere technologies to unleash their creativity through playful learning. If you've ever browsed in a toy store, you've probably seen a number of inventions that had their origins in the group's research. One is the *programmable brick technology* behind the popular LEGO Mindstorms robotics construction kits, which took their name from the landmark book *Mindstorms: Children, Computers, and Powerful Ideas* by Media Lab Professor Seymour Papert. Cognizant of the fact that LEGO Mindstorms had developed a predominantly male following, the group sought to find more ways to engage girls as well as boys and developed the *Cricket*—available commercially today as the *PicoCricket*—a high-tech arts and crafts kit that enables its users to make things such as interactive jewelry. Continuing their quest to make creators and inventors out of everyone, Resnick and his students once again collaborated with LEGO to produce the WeDo, which is about half the price of Mindstorms and is targeted at primary school students.

Another major thrust of the group's research dates back to 1993 when Resnick and his collaborator Natalie Rusk partnered with the Boston Computer Museum to cofound the *Computer Clubhouse,* an international network of afterschool centers that offer young people in low-income communities the opportunity to learn about and explore new technologies. One objective was to enable children, most of whom had never used a computer, to master the basic skills of digital technologies by providing access to an array of hardware and software tools. But even more important, the Computer Clubhouse provided children with mentoring and a supportive environment where they could learn to express themselves creatively and develop an appreciation for the benefits of collaboration and community. The program has evolved over the years, in step with the dramatic advances in technology, and today (thanks to the financial support of the Intel Corporation) there are about 100 Computer Clubhouses across the globe serving 20,000 children in twenty countries.

By 2003, each time he walked into a clubhouse, Resnick was likely to see children deftly manipulating all kinds of rich media, including graphics, animation, videos, and music, reflecting just how rapidly and widely young people had begun to embrace new digital technologies. Pundits were beginning to refer to this generation of children as "digital natives," as they appeared to have practically been born fluent in technology. But Resnick saw it differently. In his mind, true digital fluency requires not just the ability to use computers but also the ability to program them. As he puts it, "Simply using computers is like knowing how to read and not knowing how to write. In our society, you need to know how to do both."

So the group conceived and built Scratch, a programming language designed specifically to allow children to become *creators* of the kind of interactive media that they were already consuming. Children can use Scratch to mix photos, music, videos, and animations together and program everything from movies to interactive birthday cards to video games to comic books to a daily newscast, to whatever else their imagination leads them.

Scratch wasn't the first programming language designed especially for children, however. In the 1970s, Professor Seymour Papert made the first attempt to engage children in programming with a language called LOGO. Many schools experimented with LOGO in the 1980s, but the enthusiasm waned. Programming remained a narrow, technical activity until three decades later when, in true Media Lab fashion, Papert's protégé Resnick and his team made a breakthrough in programming technology, inspired by a seemingly unrelated activity: the way children play and build with LEGO blocks.

The team recalled in a recent paper: "Given a box full of them, they immediately start tinkering, snapping together a few blocks, and the emerging structure then gives them new ideas. As they play and build, plans and goals evolve organically, along with the structures. We wanted the process of programming in Scratch to have a similar feel." Based on this idea, the team built the Scratch language around *programming blocks:* predefined chunks of code that accomplish a variety of basic tasks for manipulating rich media on the screen. For example, a block might cause a Ping-Pong paddle to move up and down. The visual interface is designed to feel tangible, just like actual LEGOs, so that the programming blocks appear as various colored shapes that are selected from a palette, then dragged and dropped into an assembly area on the screen. There, they can be *snapped together* to create stacks of blocks called *scripts* that perform more complex tasks, such as controlling where the ball goes, and how fast, when it is hit by a paddle. By simply snapping lots of blocks together in this intuitive fashion, children can create entire programs—like a new type of paddle game—without having to write individual lines of code or to learn all the obscure syntax and punctuation that make programming so intimidating to many. Children with no programming experience at all can get started immediately.

Most important, they can learn to do it in a fun, playful, and "tinkerable" way. Just as when you have a box of LEGO bricks, you're not sure what you're going to make until you start putting them together, with Scratch, you never know what kind of program you'll end up with until you start trying lots of combinations. In that way, Scratch isn't just giving children the ability to express their inner programmer. It is also helping them learn the process of *iterative invention*—the same one we practice at the Media Lab. And indeed, Resnick has witnessed children tinkering in this way actually produce interactive media programs intricate enough to rival those of professional programmers.

Even the choice of the name *Scratch* reflects this approach to learning and inventing. It comes from *scratching*, a technique developed by hip-hop disc jockeys to make new sounds and rhythms from existing ones by moving a vinyl record back and forth on a turntable. The group chose the name because they felt it evoked the style of creativity—making something completely new and original by tinkering with existing forms—they sought to unleash.

Scratch was first introduced in 2005 into several Computer Clubhouse locations for testing, one at a storefront location in South Central Los Angeles, where youths from eight to eighteen came from some of the most impoverished areas of the city. The results over an eighteen-month period were amazing, with an even mix of boys and girls creating more than 1,500 interactive media programs such as games, animations, and stories. Encouraged by this, Resnick's doctoral student Karen Brennan teamed with an afterschool program for middle school students called *Citizen Schools*, through which they explored Scratch's potential as a medium for self-expression and self-understanding in a socioeconomically disadvantaged neighborhood in Boston. The children created a wide variety of projects that answered questions like *Who am I?* and *What is important to me?* Once again, the results were amazing. They validated Brennan's hypothesis that Scratch was even more than a programming language—the process of creating with it had the extraordinary effect of breaking down even longstanding emotional and social barriers among the children.

Energized by these early successes, the team then set an ambitious goal. They wanted to make Scratch available to everyone in the world. To do that, they decided to produce a website where children could share Scratch projects with one another and that would be open to everyone for free. So in spring of 2007, led by Resnick's doctoral student Andres Monroy-Hernandez, they launched http://scratch.mit.edu, an online community where children of all ages and from anywhere in the world could come together and experiment with interactive media projects like movies or music videos or video games.

But the Scratch website isn't just a platform for people to interact with the media; even more importantly, it's a place to interact and collaborate with each other as well. Users can post and share their projects with everyone, just as they would share videos on YouTube and photos on Flickr, and just like on those websites, anyone can comment on another's project. What's more, anyone can get access to the actual Scratch scripts behind other people's projects, so they can remix them, modify them, or build upon them in any way they please. To promote the spirit of collaboration, such projects point back to the original authors to give them proper credit.

Since it was introduced in 2007, the Scratch community has grown at an extraordinary pace in terms of the number members (approaching 1 million as of this writing) and the number of projects they have created (over 1.5 million as of this writing). But if you browse the Scratch website, you will be impressed not only by the *quantity* of what's there but also by the *level* of creativity that Scratch has unleashed. It's hard to think of a topic that's not addressed in an imaginative story, game, or animation. For just one example, during national elections, there's a flurry of edgy projects about the candidates that rival the parodies in the national media.

Also, the level of collaboration among members of the Scratch community across the globe exceeds anything imagined by its creators. For example, last year, a girl named Sarah and her younger brother decided to create a spooky Scratch project to celebrate Halloween, their favorite holiday. Sarah loves the programming part, and her brother loves to draw, but they wanted some help. So they posted a message on the

Scratch forums inviting others to participate in the creation of a project. One Scratcher suggested creating a game where the player has to navigate a spooky old mansion. Sarah and her brother loved the idea so the three of them started working on the plot of the story, created an initial draft of the project, which they named *Spooky Mansion,* and posted a link to the project in the forum thread. Soon, some twenty Scratchers from around the globe volunteered to help out, some with writing the plot, others with the programming, others with the art. Then, they decided to form an online "company" that they call *Blue Elk Productions,* devoted to iteratively building up the project. On the day before Halloween, Blue Elk unveiled the final version of Spooky Mansion, and it was an instant Scratch hit.

As simple as it might seem, I think there is something very profound about what Sarah and her brother and their Blue Elk Productions colleagues (and almost a million other children) are doing with Scratch. They are interacting with others in a much deeper and more creative way than they would on today's social media sites like Facebook or Twitter. When asked why she liked Scratch, Sarah declared, "I don't like to just talk with other people online. I like to talk about something creative and new." Resnick sees it this way: The ultimate goal for Scratch goes beyond computers and beyond media. As people create and share projects with Scratch, they will start to see themselves as creators, capable of contributing in important ways to the world around them. They will see their role in the world differently, and they will see themselves differently.

FURTHER REFLECTIONS

Three years after the performance at TED 2008, Dan Ellsey has become a highly sought after public speaker. He is in such demand, his calendar often double-booked, that he periodically has to cancel speaking engagements. Using his head-mounted laser pointer to manipulate text on a screen, which is then converted to synthetic speech, he addresses audiences at neurology discussion groups, graduate programs in education, nursing colleges, occupational therapy degree programs, hospitals, and preschools. He's also become a music teacher, with students coming into the hospital several times a week to collaborate with the maestro on Hyperscore.

Ellen McManus is an arts therapist at Tewksbury State Hospital who knew Ellsey long before he was introduced to Hyperscore. "Things are very different now," she says. "Dan has a very different sense of his own identity. He has a certain sense of himself that has been greatly deepened and enriched."

When I heard this, it made me think. It's hard to pinpoint when today's digital age began, but I would put it somewhere between the early 1960s (when IBM introduced the first "modern mainframe") and the early 1970s (when Intel shipped the first "modern microprocessors"). This overlaps with the lifetime of the Apollo moon program, a time when we believed that any challenge, no matter how big or audacious, was possible to achieve. That was our identity as a society. Now, despite unbelievable advances in technology that have impacted our lives in so many ways, the big challenges of our day—healthcare, education, and the environment—seem insurmountable. Sadly, that's our new identity.

But the story of how Dan Ellsey's identity was transformed is a beacon of hope for the future. Technology fully unleashed the creative genius within him, despite his so-called disabilities. There will be nine billion people on the planet by 2045, each with his or her own unique abilities and disabilities. What if future technologies such as those you read about in this book could unleash the creative genius within every one of them? Turn him or her into a musician or an engineer or a programmer? Or a doctor or a chef or a scientist? What would our identity as a society be then? I bet it would be greatly deepened and enriched, just like Dan Ellsey's. Perhaps then we would have finally fulfilled Oliver Sacks's urgent request to the audience that May morning in 2007 in the Kresge Auditorium at MIT to "humanize technology before it dehumanizes us."

Acknowledgments

The expression "I wish to thank everyone" is overused, but in the case of the Media Lab, it is literally true that *everyone* who has been there has contributed in some way to the stories told in this book. This includes all faculty members, students, staff, sponsors, and administrators; and since that numbers in the thousands, I will begin by acknowledging all of them.

I would like to express my deep gratitude to the extraordinary group of people who comprised the faculty and research staff of the Media Lab during my tenure there: Michael Bove, Ed Boyden*, Leah Buechley*, Chris Csikszentmihályi, Glorianna Davenport, Judith Donath, Matthew Goodwin*, Hugh Herr*, César Hildalgo, Henry Holtzman, Hiroshi Ishii, Joe Jacobson, Rana el Kaliouby*, Kent Larson, Henry Lieberman*, Andy Lippman, Tod Machover*, John Maeda, Pattie Maes*, Marvin Minsky, the late William Mitchell*, Nicholas Negroponte, Neri Oxman, Seymour Papert, Joe Paradiso*, Sandy Pentland*, Rosalind Picard*, Ramesh Raskar*, Mitchel Resnick*, Deb Roy*, Chris Schmandt, Ted Selker, the late Push Singh, David Small, and Barry

Vercoe. The individuals whose names are followed by an asterisk were also very generous with their time and thoughts during interviews and the editing process.

Special thanks also to the members of the "partners committee"—Hiroshi Ishii, Andy Lippman, Pattie Maes, and Mitch Resnick—who shared in the inglorious task of helping to manage this intentionally unmanageable place. John Maeda, now president of the Rhode Island School of Design, bravely served as my first associate director. And I would like to express special gratitude to my friends and confidants Tod Machover, Deb Roy, and John Hockenberry, all of whom have been a constant source of support, encouragement, perspective, and, most important, good humor during my tenure and as I wrote this book.

The creativity, energy, and enthusiasm of the students of the Media Lab are unequaled in any academic or research institution of the world, and their prodigious flow of ideas and inventions are the fuel that powers the Lab. It would be impractical to mention them all here, but I would like to give a special thank-you to those current and past students and postdoctoral fellows who were generous with their time, thoughts, and insights during the research and writing of this book: Barbara Barry, Matt Berlin, Adam Boulanger, Karen Brennan, Hazel Briner, Sonia Chernova, Ryan Chin, Angela Chang, Jae-Woo Chung, Riley Crane, Idit Harel Caperton, Shaundra Bryant Daily, Charlie DeTar, Wen Dong, Micah Eckhardt, Eran Egozy, Grant Elliott, Greg Elliott, Ian Eslick, Michael Fleischman, Catherine Havasi, Seth Hunter, Cory Kidd, Taemie Kim, Jean-Baptiste Labrune, Mat Laibowitz, Jun Ki Lee, Kwan Lee, Andy Marecki, Ernesto Martinez-Villalpando, Daniel McDuff, Pranav Mistry, Ankit Mohan, John Moore, Daniel Olguin,

Vitor Pamplona, Galen Pickard Ming-Zer Poh, Alex Rigopolous, Sajid Sadi, Peter Schmitt, Jay Silver, Cati Vaucelle, Ben Waber, Jamie Zigelbaum, and Aaron Zinman.

A very special partner in this project from its inception, Carol Colman, had not set foot in the Media Lab before we met a few years ago but quickly became a member of the family, hanging around workshops and a regular at sponsor weeks. Her friendly personality coupled with a deep intelligence and curiosity served her well in conducting the core interviews with faculty and students, and in creating drafts of the many stories. The project turned out to be much larger than either of us anticipated, and I am deeply grateful for her courage and persistence.

Nothing would have been possible for me at the Media Lab without my executive assistant Tesha Myers, who in addition to running the director's office and keeping me on track, pitched in at every stage of this book project, spanning three years from conception to completion. She was calm in the face of many challenges and was always exceedingly generous with her time and sunny spirit.

My thanks as well to all of the intrepid and overworked Media Lab staff during my tenure, with special thanks to Paula Aguilera, Ramona Allen, Paula Anzer, Lisa Breede (who was my guardian angel from the moment I arrived at the Media Lab), Amna Carreiro, John DiFrancesco, Felice Gardner, Ellen Hoffman, Henry Holtzman, Alexandra Kahn, Mary Markel Murphy, Mauro Nunez, Linda Peterson, Peter Pflanz, Greg Tucker, and Mary Young.

I am also grateful to colleagues in organizations engaged in research collaborations with the Media Lab who have contributed to the stories of creativity and invention told in the book, including Dr. George

Dimitri (director, center for Sarcoma and Bone Oncology, Dana Farber Cancer Institute), Dan Ellsey (resident and composer/conductor at the Tewksbury State Hospital), Amy Farber (founder and CEO, LAM Treatment Alliance), and Ellen McManus (arts therapist at the Tewksbury State Hospital).

The creative freedom described in this book is possible because of the generosity of the Media Lab's corporate sponsors with their resources, time, and intellectual capital. I am appreciative to them all, but a special shout out to my good friends Julius Akinyemi (formerly of PepsiCo), Joe Branc (Steelcase), Carl Hayne (P&G), Phil London of Schneider Electric, and Jeff Patmore and Steve Whittaker (BT).

I'd also like to thank Mel Berger of William Morris, who believed in this project from the first conversation and was instrumental in making it happen; Talia Krohn and Roger Scholl, my editors at Crown, who were equally strong believers and who led me through every step of the process with great patience and skill; and the publicist team of Dennelle Catlett of Crown and Sandi Mendelson of Helsinger-Mendelson.

Thanks also to a group of longtime Boston friends and colleagues who have most generously helped me express my own stories over the years as well as for this book: Andy Palmer, who is like a brother in life and in business; Janice L. Brown, my dear friend, trusted adviser, and partner in more adventures and mis-adventures than either of us would like to admit; Michael Kolowich of DigiNovations and KnowledgeVision Systems; and Micho Spring of Weber Shandwick. And thanks to a new Boston friend and colleague, Steve Bennett of AuthorBytes, for his enthusiasm and skill in building the website.

A very creative group of photographers have captured stunning images of the Media Lab and its people over the years and have contributed to those in this book, including Webb Chappell, L. Barry Hetherington, Andy Ryan, and Sam Ogden.

Lastly, my love and thanks to Mom, Kim, Ilan, Risa, and Hilary for their continual encouragement and unwavering support throughout my career, my tenure at the Media Lab, and during the course of this project. All of them offered timely thoughts and suggestions that helped me bust through the many writer's blocks that I encountered along the way. Finally, for the inspiration for telling stories and for finding the humor in just about everything, thank you, Dad.

Index